Two Manuals Combined into One Book

Military First Aid

and

Tactical Employment of Nonlethal Weapons

Department of Defense

Pentagon
Publishing

*FIELD MANUAL
NO. 4-25.11
NAVY TACTICAL
REFERENCE
PUBLICATION
NO. 4-02.1
AIR FORCE MANUAL
NO. 44-163(I)

HEADQUARTERS
DEPARTMENT OF THE ARMY,
THE NAVY, AND THE AIR FORCE
Washington, DC, 23 December 2002

FIRST AID

TABLE OF CONTENTS

*This publication supersedes FM 21-11, 27 October 1988

Page

PREFACE

This manual meets the first aid training needs of individual service members. Because medical personnel will not always be readily available, the nonmedical service members must rely heavily on their own skills and knowledge of life-sustaining methods to survive on the integrated battlefield. This publication outlines both self-aid and aid to other service members (buddy aid). More importantly, it emphasizes prompt and effective action in sustaining life and preventing or minimizing further suffering and disability. First aid is the emergency care given to the sick, injured, or wounded before being treated by medical personnel. The term *first aid* can be defined as "urgent and immediate lifesaving and other measures, which can be performed for casualties by nonmedical personnel when medical personnel are not immediately available." Nonmedical service members have received basic first aid training and should remain skilled in the correct procedures for giving first aid. This manual is directed to *all* service members. The procedures discussed apply to all types of casualties and the measures described are for use by both male and female service members.

This publication is in consonance with the following North Atlantic Treaty Organization (NATO) International Standardization Agreements (STANAGs) and American, British. Canadian, and Australian Quadripartite Standardization Agreements (QSTAGs).

TITLE	STANAG	QSTAG
Medical Training in First Aid, Basic Hygiene and Emergency Care	2122	535
First Aid Kits and Emergency Medical Care Kits	2126	
Medical First Aid and Hygiene Training in NBC Operations	2358	
First Aid Material for Chemical Injuries	2871	

These agreements are available on request, using Department of Defense (DD) Form 1425 from the Standardization Documents Order Desk, 700 Robins Avenue, Building 4, Section D, Philadelphia, Pennsylvania 19111-5094.

Unless this publication states otherwise, masculine nouns and pronouns do not refer exclusively to men.

Use of trade or brand names in this publication is for illustrative purposes only and does not imply endorsement by the Department of Defense (DOD).

The proponent for this publication is the US Army Medical Department Center and School. Submit comments and recommendations for the improvement of this publication directly to the **Commander, US Army Medical Department Center and School, ATTN: MCCS-FCD-L, 1400 East Grayson Street, Fort Sam Houston, Texas 78234-5052**.

CHAPTER 1

FUNDAMENTAL CRITERIA FOR FIRST AID

"The fate of the wounded rests in the hands of the ones who apply the first dressing."

Nicholas Senn (1898) (49th President of the American Medical Association)

1-1. General

When a nonmedical service member comes upon an unconscious or injured service member, he must accurately evaluate the casualty to determine the first aid measures needed to prevent further injury or death. He should seek medical assistance as soon as possible, but he should not interrupt the performance of first aid measures. To interrupt the first aid measures may cause more harm than good to the casualty. Remember that in a chemical environment, the service member should not evaluate the casualty until the casualty has been masked. After performing first aid, the service member must proceed with the evaluation and continue to monitor the casualty for development of conditions which may require the performance of necessary basic lifesaving measures, such as clearing the airway, rescue breathing, preventing shock, and controlling bleeding. He should continue to monitor the casualty until relieved by medical personnel.

Service members may have to depend upon their first aid knowledge and skills to save themselves (self-aid) or other service members (buddy aid/ combat lifesaver). They may be able to save a life, prevent permanent disability, or reduce long periods of hospitalization by knowing **WHAT** to do, **WHAT NOT** to do, and **WHEN** to seek medical assistance.

NOTE

The prevalence of various body armor systems currently fielded to US service members, and those in development for future fielding, may present a temporary obstacle to effective evaluation of an injured service member. You may have to *carefully remove* the body armor from the injured service member to complete the evaluation or administer first aid. Begin by removing the outer-most hard or soft body armor components (open, unfasten or cut the closures, fasteners, or straps), then remove any successive layers in the same manner. Be sure to follow other notes, cautions and warnings regarding procedures in contaminated situations and when a broken back or neck is suspected. Continue to evaluate.

1-2. Terminology

To enhance the understanding of the material contained in this publication, the following terms are used—

- *Combat lifesaver.* This is a US Army program governed by Army Regulation (AR) 350-41. The combat lifesaver is a member of a nonmedical unit selected by the unit commander for additional training beyond basic first aid procedures (referred to as *enhanced first aid*). A minimum of one individual per squad, crew, team, or equivalent-sized unit should be trained. The primary duty of this individual does not change. The additional duty of combat lifesaver is to provide enhanced first aid for injuries based on his training before the trauma specialist (military occupational specialty [MOS] 91W) arrives. The combat lifesaver's training is normally provided by medical personnel assigned, attached, or in direct support (DS) of the unit. The senior medical person designated by the commander manages the training program.

- *Trauma Specialist (US Army) or Hospital Corpsman (HM).* A medical specialist trained in emergency medical treatment (EMT) procedures and assigned or attached in support of a combat or combat support unit or marine forces.

- *Casualty evacuation.* Casualty evacuation (CASEVAC) is a term used by nonmedical units to refer to the movement of casualties aboard nonmedical vehicles or aircraft. See also the term *transported* below. Refer to FM 8-10-6 for additional information.

> **CAUTION**
>
> Casualties transported in this manner do not receive en route medical care.

- *Enhanced first aid (US Army).* Enhanced first aid is administered by the combat lifesaver. It includes measures, which require an additional level of training above self-aid and buddy aid, such as the initiation of intravenous (IV) fluids.

- *Medical evacuation.* Medical evacuation is the timely, efficient movement of the wounded, injured, or ill service members from the battlefield and other locations to medical treatment facilities (MTFs). Medical personnel provide en route medical care during the evacuation. Once the casualty has entered the medical stream (trauma specialist, hospital corpsman, evacuation crew, or MTF), the role of first aid in the care of the casualty ceases and the casualty becomes the responsibility of the health service support (HSS) chain. Once he has entered the HSS chain he is referred to as a *patient*.

- *First aid measures*. Urgent and immediate lifesaving and other measures, which can be performed for casualties (or performed by the casualty himself) by nonmedical personnel when medical personnel are not immediately available.

- *Medical treatment*. Medical treatment is the care and management of wounded, injured, or ill service members by medically trained (MOS-trained) HM, and area of concentration (AOC) personnel. It may include EMT, advanced trauma management (ATM), and resuscitative and surgical intervention.

- *Medical treatment facility*. Any facility established for the purpose of providing medical treatment. This includes battalion aid stations, Level II facilities, dispensaries, clinics, and hospitals.

- *Self-aid/buddy aid.* Each individual service member is trained to be proficient in a variety of specific first aid procedures. This training enables the service member or a buddy to apply immediate first aid measures to alleviate a life-threatening situation.

- *Transported.* A casualty is moved to an MTF in a nonmedical vehicle without en route care provided by a medically-trained service member (such as a Trauma Specialist or HM). First aid measures should be continually performed while the casualty is being transported. If the casualty is acquired by a dedicated medical vehicle with a medically-trained crew, the role of first aid ceases and the casualty becomes the responsibility of the HSS chain, and is then referred to as a *patient*. This method of transporting a casualty is also referred to as *CASEVAC*.

1-3. Understanding Vital Body Functions for First Aid

In order for the service member to learn to perform first aid procedures, he must have a basic understanding of what the vital body functions are and what the result will be if they are damaged or not functioning.

a. *Breathing Process*. All humans must have oxygen to live. Through the breathing process, the lungs draw oxygen from the air and put it into the blood. The heart pumps the blood through the body to be used by the cells that require a constant supply of oxygen. Some cells are more dependent on a constant supply of oxygen than others. For example, cells of the brain may die within 4 to 6 minutes without oxygen. Once these cells die, they are lost forever since they do not regenerate. This could result in permanent brain damage, paralysis, or death.

b. *Respiration.* Respiration occurs when a person inhales (oxygen is taken into the body) and then exhales (carbon dioxide [CO_2] is expelled from the body). Respiration involves the—

- *Airway.* The airway consists of the nose, mouth, throat, voice box, and windpipe. It is the canal through which air passes to and from the lungs.

- *Lungs.* The lungs are two elastic organs made up of thousands of tiny air spaces and covered by an airtight membrane. The *bronchial tree* is a part of the lungs.

- *Rib cage.* The rib cage is formed by the muscle-connected ribs, which join the spine in back, and the breastbone in front. The top part of the rib cage is closed by the structure of the neck, and the bottom part is separated from the abdominal cavity by a large dome-shaped muscle called the *diaphragm* (Figure 1-1). The diaphragm and rib muscles, which are under the control of the respiratory center in the brain, automatically *contract* and *relax*. *Contraction* increases and *relaxation* decreases the size of the rib cage. When the rib cage increases and then decreases, the air pressure in the lungs is first less and then more than the atmospheric pressure, thus causing the air to rush into and out of the lungs to equalize the pressure. This cycle of inhaling and exhaling is repeated about 12 to 18 times per minute.

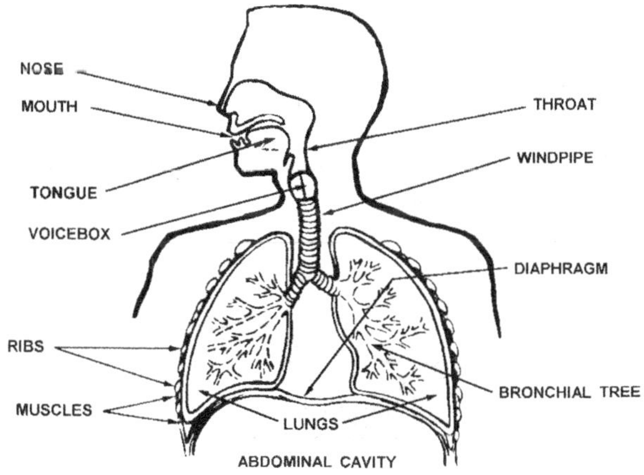

Figure 1-1. Airway, lungs, and rib cage.

c. *Blood Circulation.* The heart and the blood vessels (arteries, veins, and capillaries) circulate blood through the body tissues. The heart is divided into two separate halves, each acting as a pump. The left side pumps oxygenated blood (bright red) through the arteries into the capillaries; nutrients and oxygen pass from the blood through the walls of the capillaries into the cells. At the same time waste products and CO_2 enter the capillaries. From the capillaries the oxygen poor blood is carried through the veins to the right side of the heart and then into the lungs where it expels the CO_2 and picks up oxygen. Blood in the veins is dark red because of its low oxygen content. Blood does not flow through the veins in spurts as it does through the arteries. The entire system of the heart, blood vessels, and lymphatics is called the *circulatory system.*

(1) *Heartbeat.* The heart functions as a pump to circulate the blood continuously through the blood vessels to all parts of the body. It contracts, forcing the blood from its chambers; then it relaxes, permitting its chambers to refill with blood. The rhythmical cycle of contraction and relaxation is called the *heartbeat*. The normal heartbeat is from 60 to 80 beats per minute.

(2) *Pulse.* The heartbeat causes a rhythmical expansion and contraction of the arteries as it forces blood through them. This cycle of expansion and contraction can be felt (monitored) at various points in the body and is called the *pulse*. The common points for checking the pulse are at the—

- Side of the neck (*carotid*).
- Groin (*femoral*).
- Wrist (*radial*).
- Ankle (*posterior tibial*).

(*a*) *Carotid pulse.* To check the carotid pulse, feel for a pulse on the side of the casualty's neck closest to you. This is done by placing the tips of your first two fingers beside his Adam's apple (Figure 1-2).

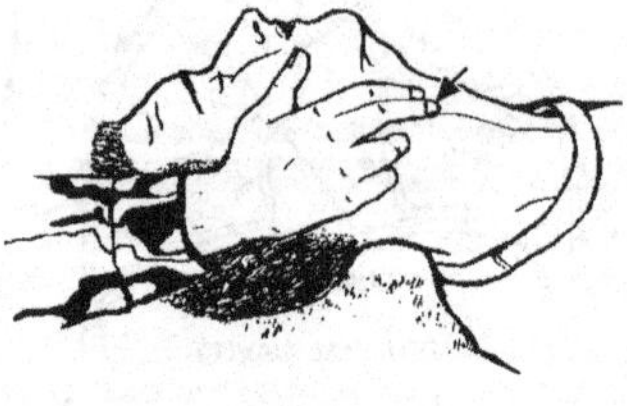

Figure 1-2. Carotid pulse.

(*b*) *Femoral pulse.* To check the femoral pulse, press the tips of your first two fingers into the middle of the groin (Figure 1-3).

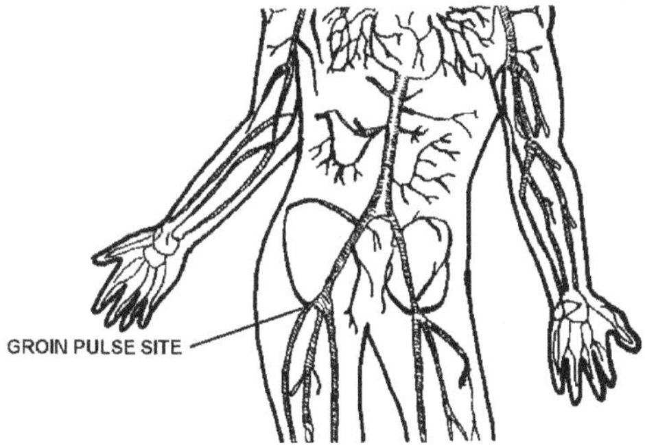

Figure 1-3. Femoral pulse.

(*c*) *Radial pulse.* To check the radial pulse, place your first two fingers on the thumb side of the casualty's wrist (Figure 1-4).

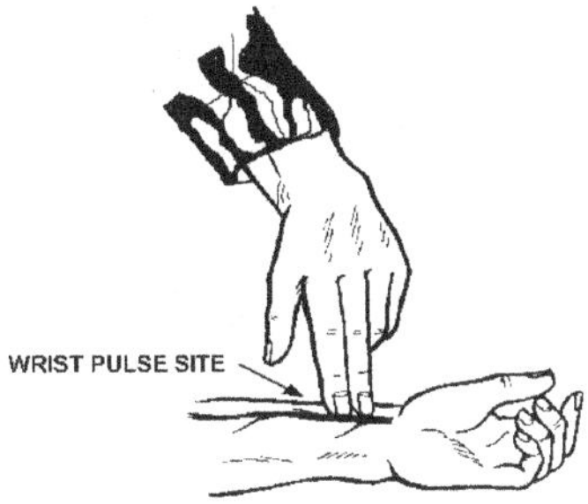

Figure 1-4. Radial pulse.

(*d*) *Posterior tibial pulse.* To check the posterior tibial pulse, place your first two fingers on the inside of the ankle (Figure 1-5).

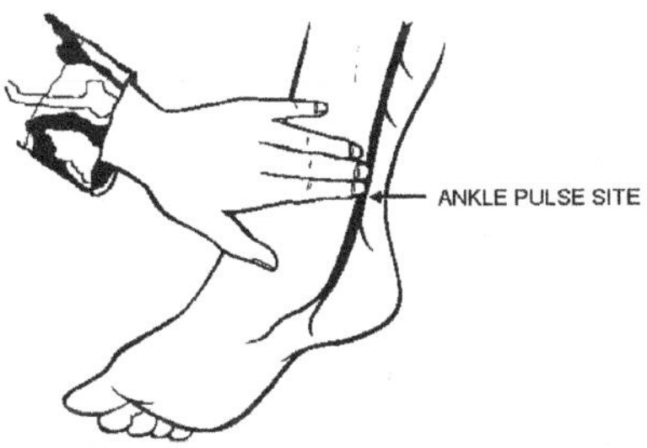

Figure 1-5. Posterior tibial pulse.

NOTE

DO NOT use your thumb to check a casualty's pulse because you may confuse the beat of your pulse with that of the casualty.

1-4. Adverse Conditions

a. *Lack of Oxygen.* Human life cannot exist without a continuous intake of oxygen. Lack of oxygen rapidly leads to death. First aid involves knowing how to open the airway and restore breathing.

b. *Bleeding.* Human life cannot continue without an adequate volume of blood circulating through the body to carry oxygen to the tissues. An important first aid measure is to stop the bleeding to prevent the loss of blood.

c. *Shock.* Shock means there is an inadequate blood flow to the vital tissues and organs. Shock that remains uncorrected may result in death even though the injury or condition causing the shock would not otherwise be fatal. Shock can result from many causes, such as loss of blood, loss of fluid from deep burns, pain, and reaction to the sight of a wound or blood. First aid includes preventing shock, since the casualty's chances of survival are much greater if he does not develop shock. Refer to paragraphs 2-21 through 2-24 for a further discussion of shock.

d. *Infection.* Recovery from a severe injury or a wound depends largely upon how well the injury or wound was initially protected. Infections result from the multiplication and growth (spread) of harmful microscopic organisms (sometimes referred to as germs). These harmful microscopic organisms are in the air, water, and soil, and on the skin and clothing. Some of these organisms will immediately invade (contaminate) a break in the skin or an open wound. The objective is to keep wounds clean and free of these organisms. A good working knowledge of basic first aid measures also includes knowing how to dress a wound to avoid infection or additional contamination.

1-5. Basics of First Aid

Most injured or ill service members are able to return to their units to fight or support primarily because they are given appropriate and timely first aid followed by the best medical care possible. Therefore, all service members must remember the basics.

- Check for **BREATHING**: Lack of oxygen intake (through a compromised airway or inadequate breathing) can lead to brain damage or death in very few minutes.

- Check for **BLEEDING**: Life cannot continue without an adequate volume of blood to carry oxygen to tissues.

- Check for **SHOCK**: Unless shock is prevented, first aid performed, and medical treatment provided, death may result even though the injury would not otherwise be fatal.

1-6. Evaluating a Casualty

a. The time may come when you must instantly apply your knowledge of first aid measures. This could occur during combat operations, in training situations, or while in a nonduty status. Any service member observing an unconscious and/or ill, injured, or wounded person must carefully and skillfully evaluate him to determine the first aid measures required to prevent further injury or death. He should seek help from medical personnel as soon as possible, but must not interrupt his evaluation of the casualty or fail to administer first aid measures. A second service member may be sent to find medical help. One of the cardinal principles for assisting a casualty is that you (the initial rescuer) must continue the evaluation and first aid measures, as the tactical situation permits, until another individual relieves you. If, during any part of the evaluation, the casualty exhibits the conditions (such as shock) for which the service member is checking, the service member must stop the evaluation and immediately administer first aid. In a chemical environment, the service member should not evaluate the casualty until both the individual and the casualty have been masked. If it is suspected that a nerve agent was used, administer the casualty's own nerve agent antidote autoinjector. After providing first aid, the service member must proceed with the evaluation and continue to monitor the casualty for further complications until relieved by medical personnel.

WARNING

Do not use your own nerve agent antidote autoinjector on the casualty.

NOTE

Remember, when evaluating and/or administering first aid to a casualty, you should seek medical aid as soon as possible. **DO NOT** stop first aid measures, but if the situation allows, send another service member to find medical aid.

b. To evaluate a casualty, perform the following steps:

(1) *Check the casualty for responsiveness.* This is done by gently shaking or tapping him while calmly asking, "Are you OK?" Watch for a response. If the casualty does not respond, go to step (2). If the casualty responds, continue with the evaluation.

(*a*) If the casualty is conscious, ask him where he feels different than usual or where it hurts. Ask him to identify the location of pain if he can, or to identify the area in which there is no feeling.

(*b*) If the casualty is conscious but is choking and cannot talk, stop the evaluation and begin first aid measures. Refer to paragraphs 2-10 and 2-11 for specific information on opening the airway.

WARNING

If a broken back or neck is suspected, do not move the casualty unless his life is in immediate danger (such as close to a burning vehicle). Movement may cause permanent paralysis or death.

(2) *Check for breathing.* (Refer to paragraph 2-6 for this procedure.)

(*a*) If the casualty is breathing, proceed to step (3).

(*b*) If the casualty is not breathing, stop the evaluation and begin first aid measures to attempt to ventilate the casualty. Attempt to open the airway, if an airway obstruction is apparent, clear the airway obstruction, then ventilate (see paragraphs 2-10 and 2-11).

(*c*) After successfully ventilating the casualty, proceed to step (3).

(3) *Check for pulse.* (Refer to paragraph 1-3*c*(2) for specific methods.) If a pulse is present and the casualty is breathing, proceed to step (4).

(*a*) If a pulse is present, but the casualty is still not breathing, start rescue breathing.

(*b*) If a pulse is not present, seek medical personnel for help.

(4) *Check for bleeding.* Look for spurts of blood or blood-soaked clothes. Also check for *both* entry and exit wounds. If the casualty is bleeding from an open wound, stop the evaluation and begin first aid procedures as follows for a—

(*a*) Wound of the arm or leg (refer to paragraphs 2-16 through 2-18 for information on putting on a field or pressure dressing).

(*b*) Partial or complete amputation, apply dressing (refer to paragraph 2-16 to 2-18) and then apply tourniquet if bleeding is not stopped (refer to paragraph 2-20 for information on putting on a tourniquet).

(*c*) Open head wound (refer to paragraph 3-10 for information on applying a dressing to an open head wound).

(*d*) Open chest wound (refer to paragraph 3-5 for information on applying a dressing to an open chest wound).

(*e*) Open abdominal wound (refer to paragraph 3-7 for information on applying a dressing to an open abdominal wound).

WARNING

In a chemically contaminated area, do not expose the wounds. Apply field dressing and then pressure dressing over wound area as needed.

(5) *Check for shock.* (Refer to paragraph 2-24 for first aid measures for shock.) If the signs and symptoms of shock are present, stop the evaluation, and begin first aid measures immediately. The following are the nine signs and symptoms of shock.

(*a*) Sweaty but cool skin (clammy skin).

(*b*) Paleness of skin. (In dark-skinned service members look for a grayish cast to the skin.)

(*c*) Restlessness or nervousness.

(*d*) Thirst.

(*e*) Loss of blood (bleeding).

(*f*) Confusion (does not seem aware of surroundings).

(*g*) Faster than normal breathing rate.

(*h*) Blotchy or bluish skin, especially around the mouth.

(*i*) Nausea or vomiting.

WARNING

Leg fractures must be splinted before elevating the legs as a first aid measure for shock.

(6) *Check for fractures*.

(*a*) Check for the following signs and symptoms of a back or neck injury and perform first aid procedures as necessary.

- Pain or tenderness of the back or neck area.
- Cuts or bruises on the back or neck area.
- Inability of a casualty to move or decreased sensation to extremities (paralysis or numbness).
  - Ask about ability to move (paralysis).
  - Touch the casualty's arms and legs and ask whether he can feel your hand (numbness).
- Unusual body or limb position.

(*b*) Immobilize any casualty suspected of having a back or neck injury by doing the following:

- Tell the casualty not to move.
- If a back injury is suspected, place padding (rolled or folded to conform to the shape of the arch) under the natural arch of the casualty's back. (For example, a blanket/poncho may be used as padding.)

> **WARNING**
>
> **Do not move casualty to place padding.**

- If a neck injury is suspected, immediately immobilize (manually) the head and neck. Place a roll of cloth under the casualty's neck, and put weighted boots (filled with dirt or sand) or rocks on both sides of his head.

(*c*) Check the casualty's arms and legs for open or closed fractures.

- Check for *open* fractures by looking for—
  - Bleeding.
  - Bones sticking through the skin.
  - Check for pulse.
- Check for *closed* fractures by looking for—
  - Swelling.
  - Discoloration.
  - Deformity.
  - Unusual body position.
  - Check for pulse.

(*d*) Stop the evaluation and begin first aid measures if a fracture to an arm or leg is suspected. Refer to Chapter 4 for information on splinting a suspected fracture.

(*e*) Check for signs/symptoms of fractures of other body areas (for example, shoulder or hip) and provide first aid as necessary.

(7) *Check for burns*. Look carefully for reddened, blistered, or charred skin; also check for singed clothing. If burns are found, stop the evaluation and begin first aid procedures. Refer to paragraph 3-9 for information on giving first aid for burns.

NOTE

Burns to the upper torso and face may cause respiratory complications. When evaluating the casualty, look for singed nose hair, soot around the nostrils, and listen for abnormal breath sounds or difficulty breathing.

(8) *Check for possible head injury.*

(*a*) Look for the following signs and symptoms:

- Unequal pupils.
- Fluid from the ear(s), nose, mouth, or injury site.
- Slurred speech.
- Confusion.
- Sleepiness.
- Loss of memory or consciousness.
- Staggering in walking.
- Headache.
- Dizziness.
- Nausea or vomiting.
- Paralysis.
- Convulsions or twitches.
- Bruising around the eyes and behind the ears.

(*b*) If a head injury is suspected, continue to watch for signs which would require performance of rescue breathing, first aid measures for shock, or control of bleeding; seek medical aid. Refer to paragraph 3-10 for information on first aid measures for head injuries.

CHAPTER 2

BASIC MEASURES FOR FIRST AID

2-1. General

Several conditions that require immediate attention are an inadequate airway, lack of breathing, and excessive loss of blood (circulation). A casualty without a clear airway or who is not breathing may die from lack of oxygen. Excessive loss of blood may lead to shock, and shock can lead to death; therefore, you must act immediately to control the loss of blood. All wounds are considered to be contaminated, since infection-producing organisms (germs) are always present on the skin and clothing, and in the soil, water, and air. Any missile or instrument (such as a bullet, shrapnel, knife, or bayonet) causing a wound pushes or carries the germs into that wound. Infection results as these organisms multiply. That a wound is contaminated does not lessen the importance of protecting it from further contamination. You must dress and bandage a wound as soon as possible to prevent further contamination.

NOTE

It is also important that you attend to any airway, breathing, or bleeding problems **IMMEDIATELY** because these problems, if left unattended, may become life threatening.

Section I. OPEN THE AIRWAY AND RESTORE BREATHING

2-2. Breathing Process

All humans must have oxygen to live. Through the breathing process, the lungs draw oxygen from the air and put it into the blood. The heart pumps the blood through the body to be used by the cells that require a constant supply of oxygen. Some cells are more dependent on a constant supply of oxygen than others. For example, cells of the brain may die within 4 to 6 minutes without oxygen. Once these cells die, they are lost forever since they do not regenerate. This could result in permanent brain damage, paralysis, or death.

2-3. Assessment of and Positioning the Casualty

a. **CHECK** for responsiveness (Figure 2-1A)—establish whether the casualty is conscious by gently shaking him and asking, "Are you OK?"

b. **CALL** for help (Figure 2-1B).

c. **POSITION** the unconscious casualty so that he is lying on his back and on a firm surface (Figure 2-1C).

> **WARNING**
>
> **If the casualty is lying on his chest (prone position), cautiously roll the casualty as a unit so that his body does not twist (which may further complicate a back, neck, or spinal injury).**

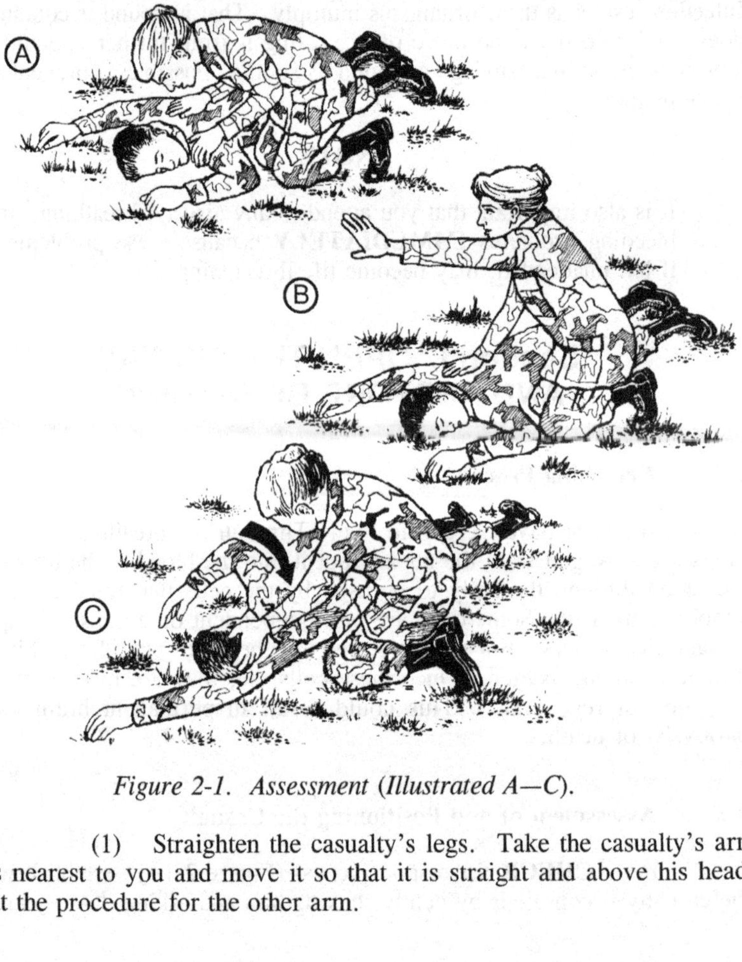

Figure 2-1. Assessment (Illustrated A—C).

(1) Straighten the casualty's legs. Take the casualty's arm that is nearest to you and move it so that it is straight and above his head. Repeat the procedure for the other arm.

(2) Kneel beside the casualty with your knees near his shoulders (leave space to roll his body) (Figure 2-1B). Place one hand behind his head and neck for support. With your other hand, grasp the casualty under his far arm (Figure 2-1C).

(3) Roll the casualty towards you using a steady, even pull. His head and neck should stay in line with his back.

(4) Return the casualty's arms to his side. Straighten his legs. Reposition yourself so that you are now kneeling at the level of the casualty's shoulders. However, if a neck injury is suspected and the jaw-thrust technique will be used, kneel at the casualty's head, looking towards his feet.

2-4. Opening the Airway of an Unconscious or Not Breathing Casualty

The tongue is the single most common cause of an airway obstruction (Figure 2-2). In most cases, simply using the head-tilt/chin-lift technique can clear the airway. This action pulls the tongue away from the air passage in the throat (Figure 2-3).

Figure 2-2. Airway blocked by tongue.

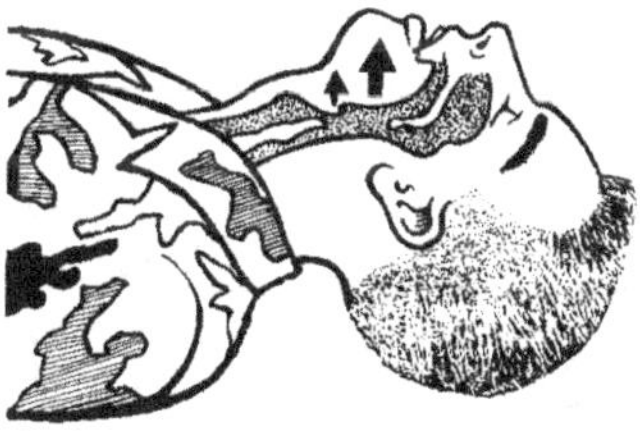

Figure 2-3. Airway opened by extending neck.

a. Call for help and then position the casualty. Move (roll) the casualty onto his back (Figure 2-1C). (Refer to paragraph 2-3*c* for information on positioning the casualty.)

NOTE

Perform finger sweep. If foreign material or vomitus is visible in the mouth, it should be removed, but do not spend an excessive amount of time doing so.

b. Open the airway using the jaw-thrust or head-tilt/chin-lift technique.

CAUTION

The head-tilt/chin-lift technique is an important procedure in opening the airway; however, use extreme care because excess force in performing this maneuver may cause further spinal injury. In a casualty with a suspected neck injury or severe head trauma, the safest approach to opening the airway is the jaw-thrust technique because in most cases it can be accomplished without extending the neck.

(1) *Perform the jaw-thrust technique.* The jaw-thrust may be accomplished by the rescuer grasping the angles of the casualty's lower jaw and lifting with both hands, one on each side, displacing the jaw forward and up (Figure 2-4). The rescuer's elbows should rest on the surface on which the casualty is lying. If the lips close, the lower lip can be retracted with the thumb. If mouth-to-mouth breathing is necessary, close the nostrils by placing your cheek tightly against them. The head should be carefully supported without tilting it backwards or turning it from side to side. If this is unsuccessful, the head should be tilted back very slightly. The jaw-thrust is the safest first approach to opening the airway of a casualty who has a suspected neck injury because in most cases it can be accomplished without extending the neck.

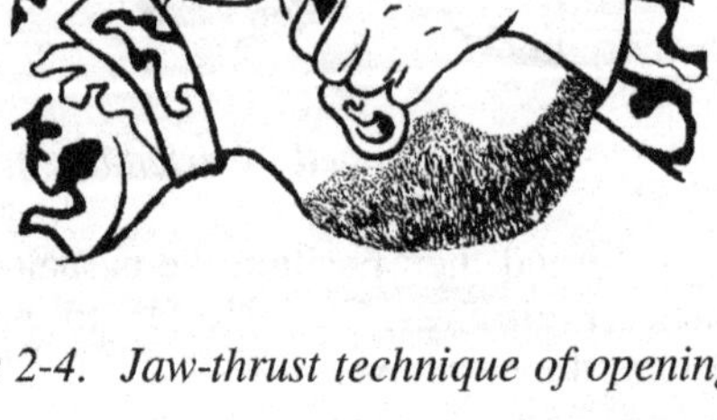

Figure 2-4. Jaw-thrust technique of opening airway.

(2) *Perform the head-tilt/chin-lift technique.* Place one hand on the casualty's forehead and apply firm, backward pressure with the palm to tilt the head back. Place the fingertips of the other hand under the bony part of the lower jaw and lift, bringing the chin forward. The thumb should not be used to lift the chin (Figure 2-5).

NOTE

The fingers should not press deeply into the soft tissue under the chin because the airway may be obstructed.

Figure 2-5. Head-tilt/chin-lift technique of opening airway.

(3) *Check for breathing (while maintaining an airway).* After establishing an open airway, it is important to maintain that airway in an open position. Often the act of just opening and maintaining the airway will allow the casualty to breathe properly. Once the rescuer uses one of the techniques to open the airway (jaw-thrust or head-tilt/chin-lift), he should maintain that head position to keep the airway open. Failure to maintain the open airway will prevent the casualty from receiving an adequate supply of oxygen. Therefore, while maintaining an open airway the rescuer should check for breathing by observing the casualty's chest and performing the following actions within 3 to 5 seconds:

(*a*) **LOOK** for the chest to rise and fall.

(*b*) **LISTEN** for air escaping during exhalation by placing your ear near the casualty's mouth.

(*c*) **FEEL** for the flow of air on your cheek (see Figure 2-6).

(*d*) **PERFORM** rescue breathing if the casualty does not resume breathing spontaneously.

NOTE

If the casualty resumes breathing, monitor and maintain the open airway. He should be transported to an MTF, as soon as practical.

2-5. Rescue Breathing (Artificial Respiration)

a. If the casualty does not promptly resume adequate spontaneous breathing after the airway is open, rescue breathing (artificial respiration) must be started. Be calm! Think and act quickly! The sooner you begin rescue breathing, the more likely you are to restore the casualty's breathing. If you are in doubt whether the casualty is breathing, give artificial respiration, since it can do no harm to a person who is breathing. If the casualty is breathing, you can feel and see his chest move. If the casualty is breathing, you can feel and hear air being expelled by putting your hand or ear close to his mouth and nose.

b. There are several methods of administering rescue breathing. The mouth-to-mouth method is preferred; however, it cannot be used in all situations. If the casualty has a severe jaw fracture or mouth wound or his jaws are tightly closed by spasms, use the mouth-to-nose method.

2-6. Preliminary Steps—All Rescue Breathing Methods

a. Establish unresponsiveness. Call for help. Turn or position the casualty.

b. Open the airway.

c. Check for breathing by placing your ear over the casualty's mouth and nose, and looking toward his chest.

(1) **LOOK** for rise and fall of the casualty's chest (Figure 2-6).

(2) **LISTEN** for sounds of breathing.

(3) **FEEL** for breath on the side of your face. If the chest does not rise and fall and no air is exhaled, then the casualty is not breathing.

(4) **PERFORM** rescue breathing if the casualty is not breathing.

NOTE

Although the rescuer may notice that the casualty is making respiratory efforts, the airway may still be obstructed and opening the airway may be all that is needed. If the casualty resumes breathing, the rescuer should continue to maintain an open airway.

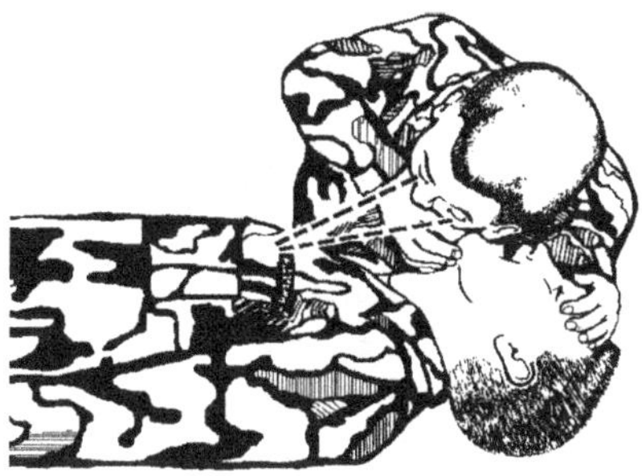

Figure 2-6. Check for breathing.

2-7. Mouth-to-Mouth Method

In this method of rescue breathing, you inflate the casualty's lungs with air from your lungs. This can be accomplished by blowing air into the person's mouth. The mouth-to-mouth rescue breathing method is performed as follows:

a. If the casualty is not breathing, place your hand on his forehead, and pinch his nostrils together with the thumb and index finger of this hand. Let this same hand exert pressure on his forehead to maintain the backward head tilt and maintain an open airway. With your other hand, keep your fingertips on the bony part of the lower jaw near the chin and lift (Figure 2-7).

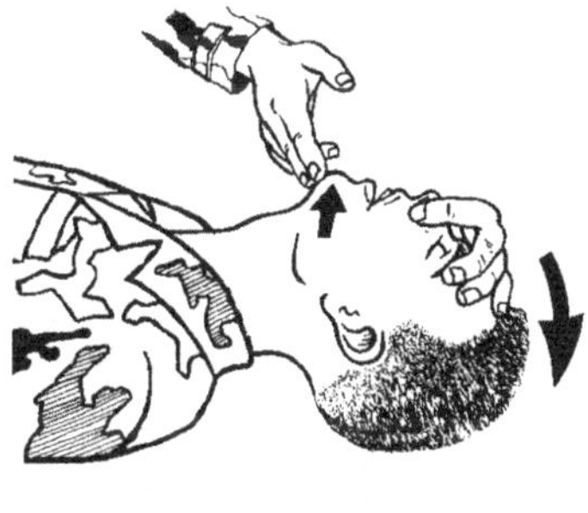

Figure 2-7. Head tilt/chin lift.

NOTE

If you suspect the casualty has a neck injury and you are using the jaw-thrust technique, close the nostrils by placing your cheek tightly against them.

b. Take a deep breath and place your mouth (in an airtight seal) around the casualty's mouth (Figure 2-8). (If the injured person is small, cover both his nose and mouth with your mouth, sealing your lips against the skin of his face.)

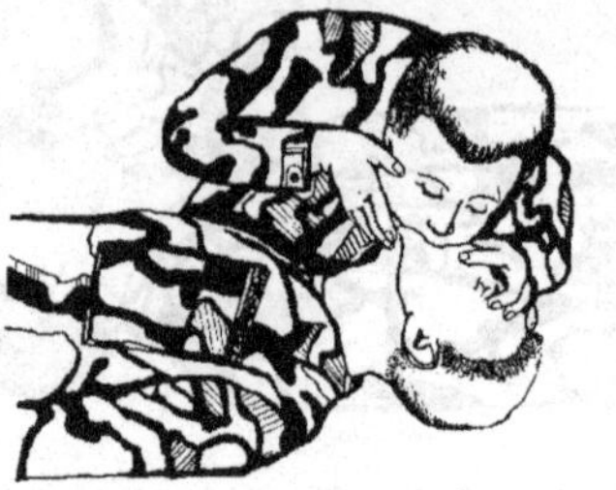

Figure 2-8. Rescue breathing.

c. Blow two full breaths into the casualty's mouth (1 to 1 1/2 seconds per breath), taking a breath of fresh air each time before you blow. Watch out of the corner of your eye for the casualty's chest to rise. If the chest rises, sufficient air is getting into the casualty's lungs. Therefore, proceed as described in step (1). If the chest does not rise, do the following (*a*, *b*, and *c* below) and then attempt to ventilate again.

(1) Take corrective action immediately by reestablishing the airway. Make sure that air is not leaking from around your mouth or out of the casualty's pinched nose.

(2) Reattempt to ventilate.

(3) If the chest still does not rise, take the necessary action to open an obstructed airway (paragraph 2-10).

NOTE

If the initial attempt to ventilate the casualty is unsuccessful, reposition the casualty's head and repeat rescue breathing. Improper chin and head positioning is the most common cause of difficulty with ventilation. If the casualty cannot be ventilated after repositioning the head, proceed with foreign-body airway obstruction maneuvers (see paragraph 2-10).

(4) After giving two slow breaths, which cause the chest to rise, attempt to locate a pulse on the casualty. Feel for a pulse on the side of the casualty's neck closest to you by placing the first two fingers (index and middle fingers) of your hand on the groove beside the casualty's Adam's apple (carotid pulse) (Figure 2-9). (Your thumb should not be used for pulse taking because you may confuse your pulse beat with that of the casualty.) Maintain the airway by keeping your other hand on the casualty's forehead. Allow 5 to 10 seconds to determine if there is a pulse.

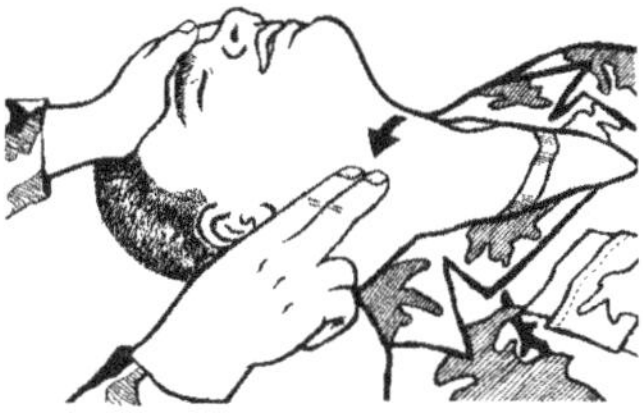

Figure 2-9. Placement of fingers to detect pulse.

(*a*) If signs of circulation are present and a pulse is found and the casualty is breathing—**STOP**; allow the casualty to breathe on his own. If possible, keep him warm and comfortable.

(*b*) If a pulse is found and the casualty is not breathing, continue rescue breathing.

(*c*) If a pulse is not found, seek medically trained personnel for help as soon as possible.

2-8. Mouth-to-Nose Method

Use this method if you cannot perform mouth-to-mouth rescue breathing because the casualty has a severe jaw fracture or mouth wound or his jaws are tightly closed by spasms. The mouth-to-nose method is performed in the same way as the mouth-to-mouth method except that you blow into the nose while you hold the lips closed with one hand at the chin. You then remove your mouth to allow the casualty to exhale passively. It may be necessary to separate the casualty's lips to allow the air to escape during exhalation.

2-9. Heartbeat

If a casualty's heart stops beating, you must immediately seek medical help. **SECONDS COUNT!** Stoppage of the heart is soon followed by cessation of respiration unless it has occurred first. Be calm! Think and act! When a casualty's heart has stopped, there is no pulse at all; the person is unconscious and limp, and the pupils of his eyes are open wide. When evaluating a casualty or when performing the preliminary steps of rescue breathing, feel for a pulse. If you DO NOT detect a pulse, seek medical help.

2-10. Airway Obstructions

In order for oxygen from the air to flow to and from the lungs, the upper airway must be unobstructed.

a. Upper airway obstructions often occur because—

(1) The casualty's tongue falls back into his throat while he is unconscious. The tongue *falls back* and *obstructs* the airway, it is not swallowed by the casualty.

NOTE

Ensure the correct positioning and maintenance of the open airway for an injured or unconscious casualty.

(2) Foreign bodies become lodged in the throat. These obstructions usually occur while eating. Choking on food (usually meat) is associated with—

- Attempting to swallow large pieces of poorly chewed food.
- Drinking alcohol.
- Slipping dentures.

(3) The contents of the stomach are regurgitated and may block the airway.

(4) Blood clots may form as a result of head and facial injuries.

b. Upper airway obstruction may cause either partial or complete airway blockage.

(1) *Partial airway obstruction.* The casualty may still have an air exchange. A good air exchange means that the casualty can cough

forcefully, though he may be wheezing between coughs. You, the rescuer, should not interfere, and should encourage the casualty to cough up the object obstructing his airway on his own. A poor air exchange may be indicated by weak coughing with a high pitched noise between coughs. Further, the casualty may show signs of shock (paragraph 1-6*b*[5]) indicating a need for oxygen. You should assist the casualty and treat him as though he had a complete obstruction.

(2) *Complete airway obstruction.* A complete obstruction (no air exchange) is indicated if the casualty cannot speak, breathe, or cough at all. He may be clutching his neck and moving erratically. In an unconscious casualty, a complete obstruction is also indicated if after opening his airway you cannot ventilate him.

2-11. Opening the Obstructed Airway—Conscious Casualty

Clearing a conscious casualty's airway obstruction can be performed with the casualty either standing or sitting and by following a relatively simple procedure.

> **WARNING**
>
> **Once an obstructed airway occurs, the brain will develop an oxygen deficiency resulting in unconsciousness. Death will follow rapidly if breathing is not promptly restored.**

a. Ask the casualty if he can speak or if he is choking. Check for the universal choking sign (Figure 2-10).

Figure 2-10. Universal sign of choking.

b. If the casualty can speak, encourage him to attempt to cough; the casualty still has a good air exchange. If he is able to speak or cough effectively, DO NOT interfere with his attempts to expel the obstruction.

c. Listen for high pitched sounds when the casualty breathes or coughs (poor air exchange). If there is poor air exchange or no breathing, CALL FOR HELP and immediately deliver manual thrusts (either an abdominal or chest thrust).

NOTE

The manual thrust with the hands centered between the waist and the rib cage is called an abdominal thrust (or Heimlich maneuver). The chest thrust (the hands are centered in the middle of the breastbone) is used only for an individual in the advanced stages of pregnancy, in the markedly obese casualty, or if there is a significant abdominal wound.

(1) Apply abdominal thrusts. This can be accomplished by using the following procedures:

(*a*) Stand behind the casualty and wrap your arms around his waist.

(*b*) Make a fist with one hand and grasp it with the other. The thumb side of your fist should be against the casualty's abdomen, in the midline and slightly above the casualty's navel, but well below the tip of the breastbone (Figure 2-11).

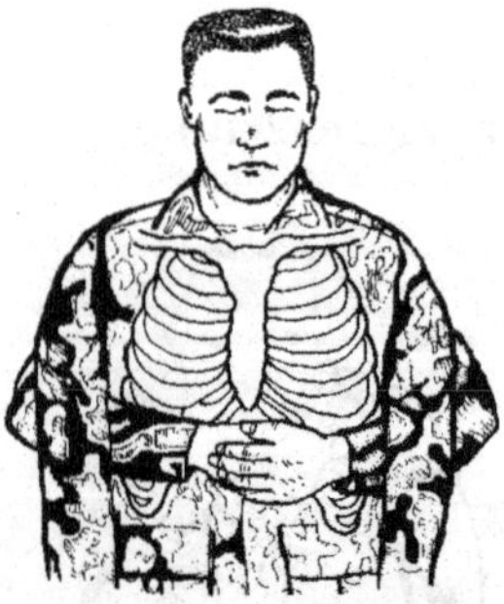

Figure 2-11. Anatomical view of abdominal thrust procedure.

(*c*) Press the fists into the abdomen with a quick backward and upward thrust (Figure 2-12).

Figure 2-12. Profile view of abdominal thrust.

(*d*) Each thrust should be a separate and distinct movement.

NOTE

Continue performing abdominal thrusts until the obstruction is expelled or the casualty becomes unresponsive.

(*e*) If the casualty becomes unresponsive, call for help as you proceed with steps to open the airway, and perform rescue breathing. (Refer to paragraph 2-7 for information on how to perform mouth-to-mouth resuscitation.)

(2) Apply chest thrusts. An alternate technique to the abdominal thrust is the chest thrust. This technique is useful when the casualty has an abdominal wound, when the casualty is pregnant, or when the casualty is so large that you cannot wrap your arms around the abdomen. To apply chest thrusts with casualty sitting or standing:

(*a*) Stand behind the casualty and wrap your arms around his chest with your arms under his armpits.

(*b*) Make a fist with one hand and place the thumb side of the fist in the middle of the breastbone (take care to avoid the tip of the breastbone and the margins of the ribs).

(*c*) Grasp the fist with the other hand and exert thrusts (Figure 2-13).

Figure 2-13. Profile view of chest thrust.

(*d*) Each thrust should be delivered slowly, distinctly, and with the intent of relieving the obstruction.

(*e*) Perform chest thrusts until the obstruction is expelled or the casualty becomes unresponsive.

(*f*) If the casualty becomes unresponsive, call for help as you proceed with steps to open the airway and perform rescue breathing.

2-12. Opening the Obstructed Airway—Casualty Lying Down or Unresponsive

The following procedures are used to expel an airway obstruction in a casualty who is lying down, who becomes unconscious, or who is found unconscious (the cause unknown):

- If a conscious casualty who is choking becomes unresponsive, call for help, open the airway, perform a finger sweep, and attempt rescue breathing (paragraphs 2-4 through 2-8). If you still cannot administer rescue breathing due to an airway blockage, then remove the airway obstruction using the procedures as in *b* below.

- If a casualty is unresponsive when you find him (the cause unknown), assess or evaluate the situation, call for help, position the casualty on his back, open the airway, establish breathlessness, and attempt to perform rescue breathing (paragraphs 2-4 through 2-8).

a. Open the airway and attempt rescue breathing (refer to paragraph 2-7 for information on how to perform mouth-to-mouth resuscitation).

b. If still unable to ventilate the casualty, perform 6 to 10 manual (abdominal or chest) thrusts.

(1) To perform the abdominal thrusts:

(*a*) Kneel astride the casualty's thighs (Figure 2-14).

Figure 2-14. Abdominal thrust on unresponsive casualty.

(*b*) Place the heel of one hand against the casualty's abdomen (in the midline slightly above the navel but well below the tip of the breastbone). Place your other hand on top of the first one. Point your fingers toward the casualty's head.

(*c*) Press into the casualty's abdomen with a quick, forward and upward thrust. You can use your body weight to perform the maneuver. Deliver each thrust quickly and distinctly.

(*d*) Repeat the sequence of abdominal thrusts, finger sweep, and rescue breathing (attempt to ventilate) as long as necessary to remove the object from the obstructed airway.

(*e*) If the casualty's chest rises, proceed to feeling for pulse.

(2) To perform chest thrusts:

(*a*) Place the unresponsive casualty on his back, face up, and open his mouth. Kneel close to the side of the casualty's body.

1. Locate the lower edge of the casualty's ribs with your fingers. Run the fingers up along the rib cage to the notch (Figure 2-15A).

2. Place the middle finger on the notch and the index finger next to the middle finger on the lower edge of the breastbone. Place the heel of the other hand on the lower half of the breastbone next to the two fingers (Figure 2-15B).

3. Remove the fingers from the notch and place that hand on top of the positioned hand on the breastbone, extending or interlocking the fingers (Figure 2-15C).

4. Straighten and lock your elbows with your shoulders directly above your hands without bending the elbows, rocking, or allowing the shoulders to sag. Apply enough pressure to depress the breastbone 1 1/2 to 2 inches, then release the pressure completely (Figure 2-15D). Do this 6 to 10 times. Each thrust should be delivered quickly and distinctly. See Figure 2-16 for another view of the breastbone being depressed.

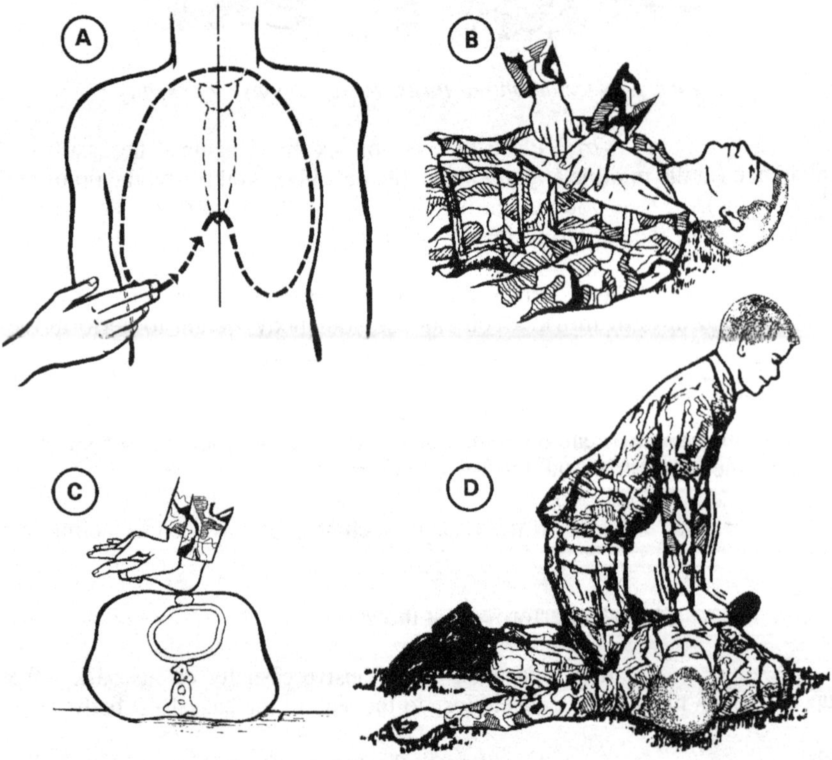

Figure 2-15. Hand placement for chest thrust (Illustrated A-D).

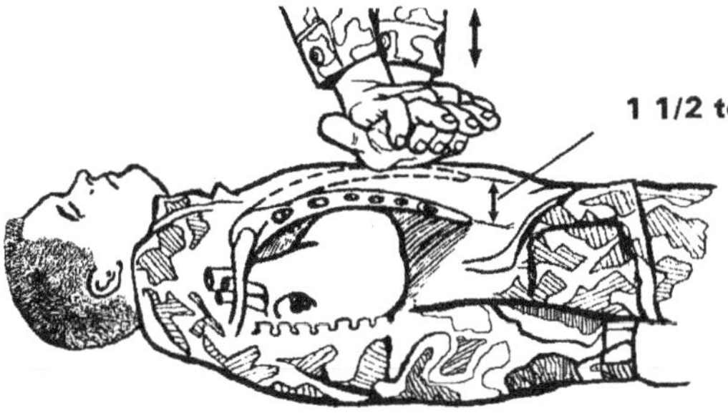

Figure 2-16. Breastbone depressed 1 1/2 to 2 inches.

(*b*) Repeat the sequence of chest thrust, finger sweep, and rescue breathing as long as necessary to clear the object from the obstructed airway. See paragraph (3) below.

(*c*) If the casualty's chest rises, proceed to feeling for his pulse.

(3) If you still cannot administer rescue breathing due to an airway obstruction, then remove the airway obstruction using the procedures in steps (*a*) and (*b*) below.

(*a*) Place the casualty on his back, face up, turn the unresponsive casualty as a unit, and call out for help.

(*b*) Perform finger sweep, keep casualty face up, use tongue-jaw lift to open mouth.

1. Open the casualty's mouth by grasping both his tongue and lower jaw between your thumb and fingers and lifting (tongue-jaw lift) (Figure 2-17). If you are unable to open his mouth, cross your fingers and thumb (crossed-finger method) and push his teeth apart (Figure 2-18) by pressing your thumb against his upper teeth and pressing your finger against his lower teeth.

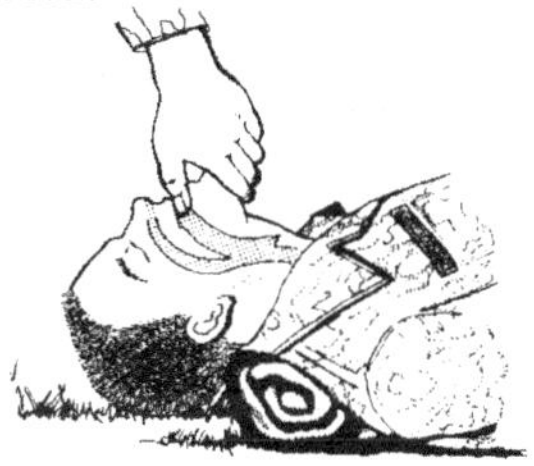

Figure 2-17. Opening casualty's mouth (tongue-jaw lift).

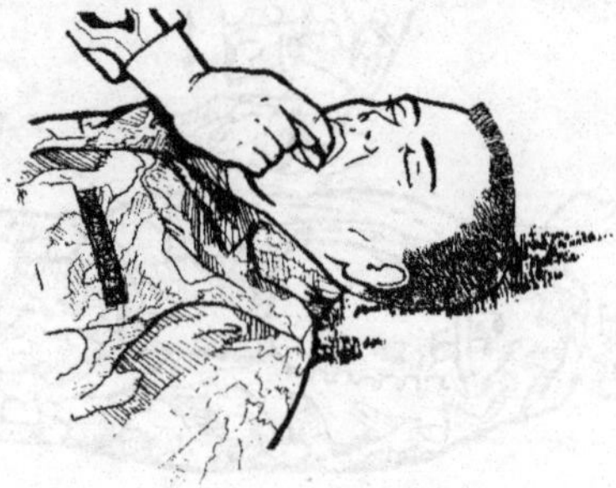

Figure 2-18. Opening casualty's mouth (crossed-finger method).

2. Insert the index finger of the other hand down along the inside of his cheek to the base of the tongue. Use a hooking motion from the side of the mouth toward the center to dislodge the foreign body (Figure 2-19).

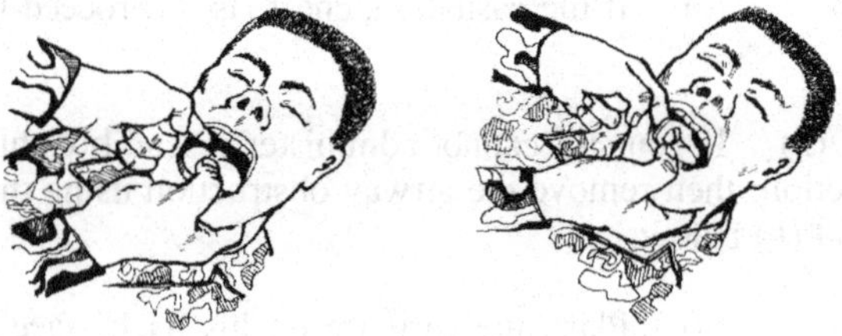

Figure 2-19. Using finger to dislodge a foreign body.

WARNING

Take care not to force the object deeper into the airway by pushing it with the finger.

Section II. STOP THE BLEEDING AND PROTECT THE WOUND

2-13. General

The longer a service member bleeds from a major wound, the less likely he will be able to survive his injuries. It is, therefore, important that the first aid provider promptly stop the external bleeding.

2-14. Clothing

In evaluating the casualty for location, type, and size of the wound or injury, cut or tear his clothing and carefully expose the entire area of the wound. This procedure is necessary to properly visualize injury and avoid further contamination. Clothing stuck to the wound should be left in place to avoid further injury. DO NOT touch the wound; keep it as clean as possible.

WARNING

DO NOT REMOVE protective clothing in a chemical environment. Apply dressings over the protective clothing.

2-15. Entrance and Exit Wounds

Before applying the dressing, carefully examine the casualty to determine if there is more than one wound. A missile may have entered at one point and exited at another point. The *EXIT* wound is usually *LARGER* than the entrance wound.

WARNING

The casualty should be continually monitored for development of conditions which may require the performance of necessary basic lifesaving measures, such as clearing the airway and mouth-to-mouth resuscitation. All open (or penetrating) wounds should be checked for a point of entry and exit and first aid measures applied accordingly.

WARNING

If the missile lodges in the body (fails to exit), DO NOT attempt to remove it or probe the wound. Apply a dressing. If there is an object extending from (impaled in) the wound, DO NOT remove the object. Apply a dressing around the object and use additional improvised bulky materials/dressings (use the cleanest material available) to build up the area around the object to stabilize the object and prevent further injury. Apply a supporting bandage over the bulky materials to hold them in place.

2-16. Field Dressing

a. Use the casualty's field dressing; remove it from the wrapper and grasp the tails of the dressing with both hands (Figure 2-20).

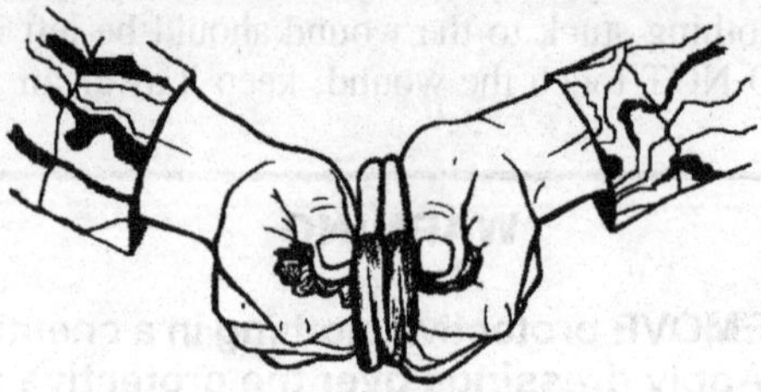

Figure 2-20. Grasping tails of dressing with both hands.

> **WARNING**
>
> **DO NOT touch the white (sterile) side of the dressing, and DO NOT allow it to come in contact with any surface other than the wound.**

b. Hold the dressing directly over the wound with the white side down. Pull the dressing open (Figure 2-21) and place it directly over the wound (Figure 2-22).

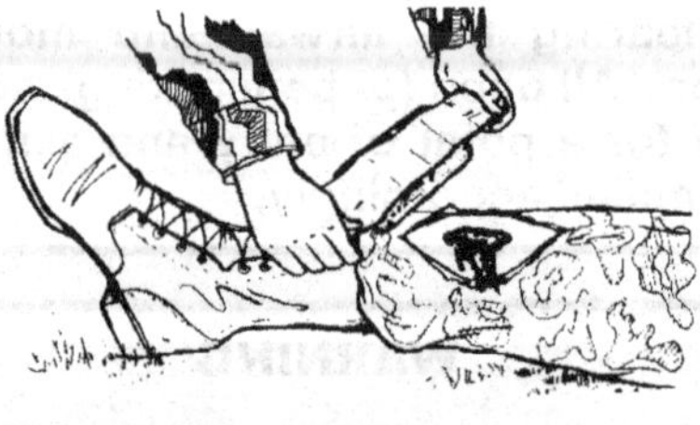

Figure 2-21. Pulling dressing open.

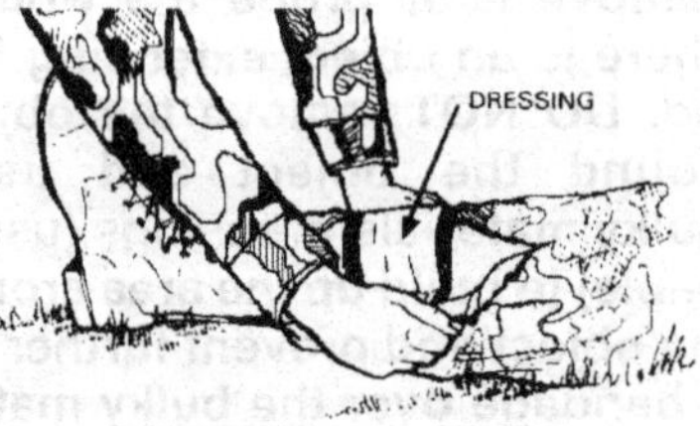

Figure 2-22. Placing dressing directly on wound.

c. Hold the dressing in place with one hand. Use the other hand to wrap one of the tails around the injured part, covering about one-half of the dressing (Figure 2-23). Leave enough of the tail for a knot. If the casualty is able, he may assist by holding the dressing in place.

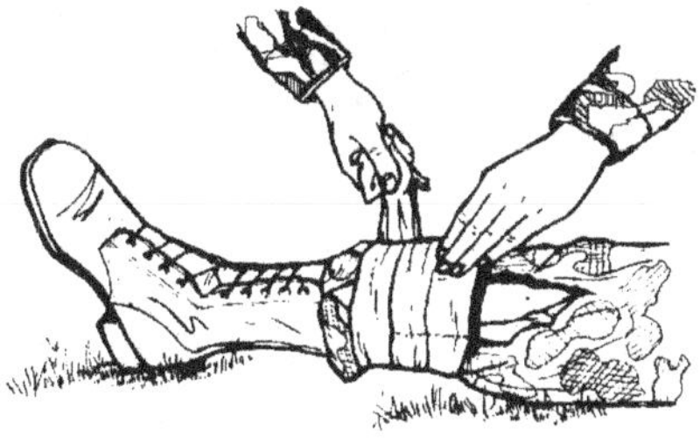

Figure 2-23. Wrapping tail of dressing around injured part.

d. Wrap the other tail in the opposite direction until the remainder of the dressing is covered. The tails should seal the sides of the dressing to keep foreign material from getting under it.

e. Tie the tails into a nonslip knot over the outer edge of the dressing (Figure 2-24). **DO NOT TIE THE KNOT OVER THE WOUND.** In order to allow blood to flow to the rest of an injured limb, tie the dressing firmly enough to prevent it from slipping but without causing a tourniquet-like effect; that is, the skin beyond the injury should not becomes cool, blue, or numb.

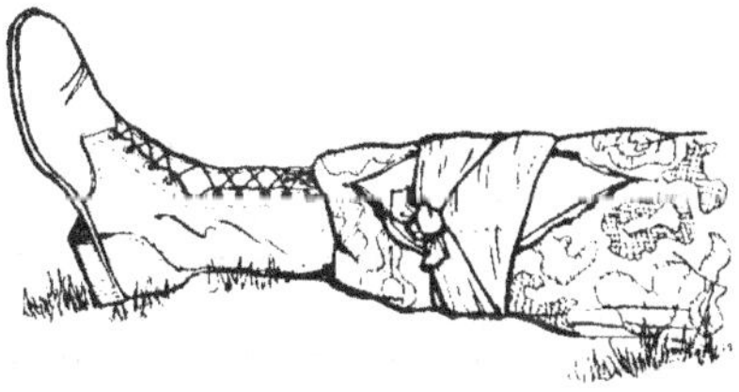

Figure 2-24. Tails tied into nonslip knot.

2-17. Manual Pressure

a. If bleeding continues after applying the sterile field dressing, direct manual pressure may be used to help control bleeding. Apply such pressure by placing a hand on the dressing and exerting firm pressure for 5 to 10 minutes (Figure 2-25). The casualty may be asked to do this himself if he is conscious and can follow instructions.

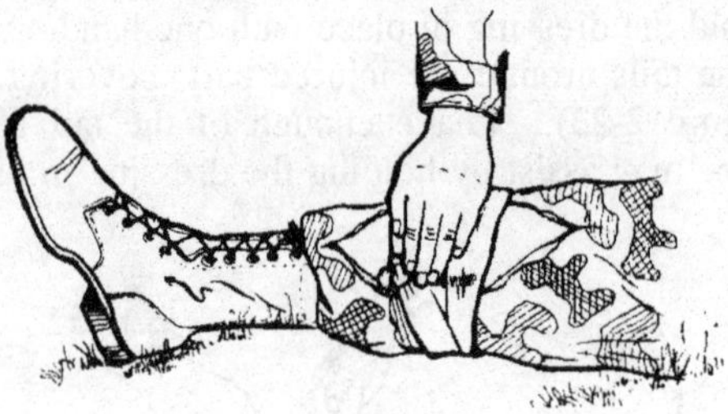

Figure 2-25. Direct manual pressure applied.

b. Elevate an injured limb slightly above the level of the heart to reduce the bleeding (Figure 2-26).

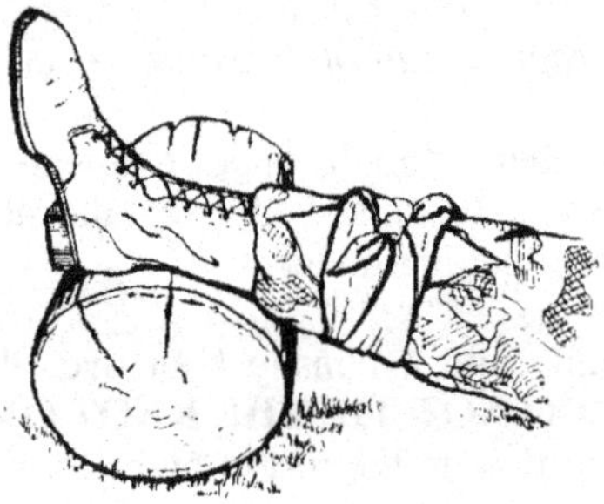

Figure 2-26. Injured limb elevated.

WARNING

DO NOT elevate a suspected fractured limb unless it has been properly splinted.

c. If the bleeding stops, check shock; administer first aid for shock as necessary. If the bleeding continues, apply a pressure dressing.

2-18. Pressure Dressing

Pressure dressings aid in blood clotting and compress the open blood vessel. If bleeding continues after the application of a field dressing, manual pressure, and elevation, then a pressure dressing must be applied as follows:

a. Place a wad of padding on top of the field dressing, directly over the wound (Figure 2-27). Keep the injured extremity elevated.

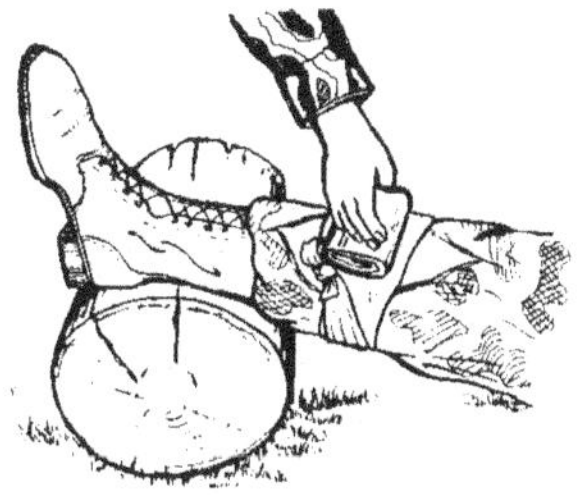

Figure 2-27. Wad of padding on top of field dressing.

NOTE

Improvised bandages may be made from strips of cloth. These strips may be made from T-shirts, socks, or other garments.

b. Place an improvised dressing (or cravat, if available) over the wad of padding (Figure 2-28). Wrap the ends tightly around the injured limb, covering the previously placed field dressing (Figure 2-29).

Figure 2-28. Improvised dressing over wad of padding

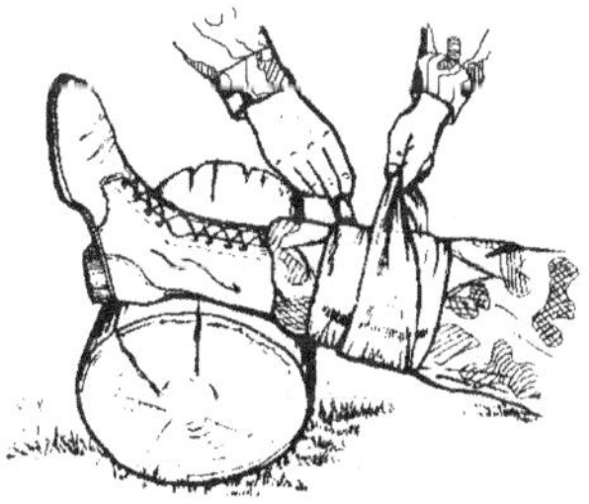

Figure 2-29. Ends of improvised dressing wrapped tightly around limb.

c. Tie the ends together in a nonslip knot, directly over the wound site (Figure 2-30). DO NOT tie so tightly that it has a tourniquet-like effect. If bleeding continues and all other measures have failed, or if the limb is severed, then apply a tourniquet. Use the tourniquet as a **LAST RESORT**. When the bleeding stops, check for shock; administer first aid for shock as necessary.

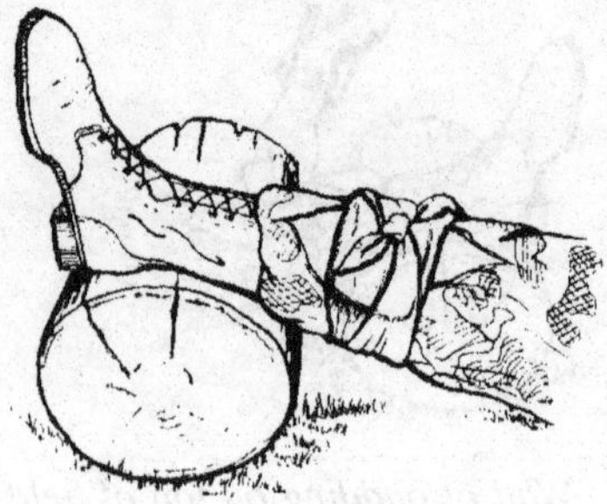

Figure 2-30. Ends of improvised dressing tied together in nonslip knot.

NOTE

Distal end of wounded extremities (fingers and toes) should be checked periodically for adequate circulation. The dressing must be loosened if the extremity becomes cool, blue, or numb.

NOTE

If bleeding continues and all other measures have failed (dressings and covering wound, applying direct manual pressure, elevating the limb above the heart level, and applying a pressure dressing while maintaining limb elevation) *then apply digital pressure* (see paragraph 2-19).

2-19. Digital Pressure

Digital pressure (often called "pressure points") is an alternative method to control bleeding. This method uses pressure from the fingers, thumbs, or hands to press at the site or point where a main artery supplying the wounded area lies near the skin surface or over bone (Figure 2-31). This pressure may help shut off or slow down the flow of blood from the heart to the wound and is used in combination with direct pressure and elevation. It may help in instances where bleeding is not easily controlled, where a pressure dressing has not yet been applied, or where pressure dressings are not readily available.

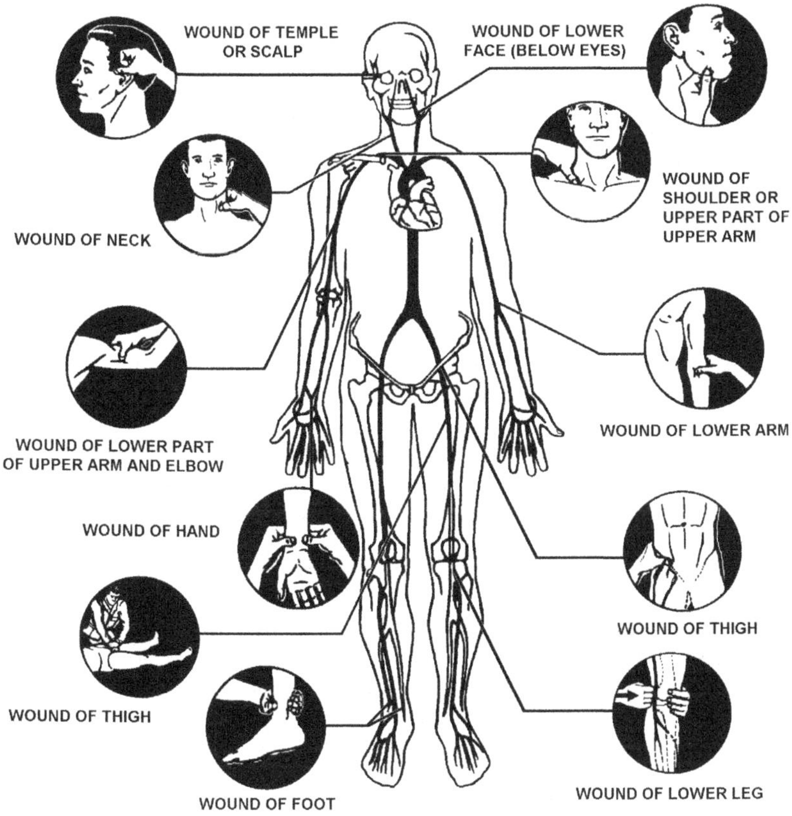

Figure 2-31. Digital pressure (pressure with fingers, thumbs or hands).

2-20. Tourniquet

DANGER

A tourniquet is only used on an arm or leg where there is a danger of the casualty losing his life (bleeding to death).

A tourniquet is a constricting band placed around an arm or leg to control bleeding. A service member whose arm or leg has been completely amputated may not be bleeding when first discovered, but a tourniquet should be applied anyway. This absence of bleeding is due to the body's normal defenses (contraction or clotting of blood vessels) as a result of the amputation, but after a period of time bleeding will start as the blood vessels relax or the clot may be knocked loose by moving the casualty. Bleeding from a major artery of the thigh, lower leg, or arm and bleeding from multiple arteries (which occurs in a traumatic amputation) may prove to be beyond control by manual pressure. If the pressure dressing (see paragraph 2-18, above) under firm hand pressure becomes soaked with blood and the wound continues to bleed, apply a tourniquet.

WARNING

Casualty should be continually monitored for development of conditions which may require the performance of necessary basic lifesaving measures, such as: clearing the airway, performing mouth-to-mouth resuscitation, preventing shock, and/or bleeding control. All open (or penetrating) wounds should be checked for a point of entry or exit and treated accordingly.

The tourniquet should not be used unless a pressure dressing has failed to stop the bleeding or an arm or leg has been cut off. On occasion, tourniquets have injured blood vessels and nerves. If left in place too long, a tourniquet can cause loss of an arm or leg. Once applied, it must stay in place, and the casualty must be taken to the nearest MTF as soon as possible. *DO NOT loosen or release a tourniquet after it has been applied as release could precipitate bleeding and potentially lead to shock.*

a. Improvising a Tourniquet. In the absence of a specially designed tourniquet, a tourniquet may be made from a strong, pliable material, such as gauze or muslin bandages, clothing, or cravats. An improvised tourniquet is used with a rigid stick-like object. To minimize skin damage, ensure that the improvised tourniquet is at least 2 inches wide.

WARNING

The tourniquet must be easily identified or easily seen.

WARNING

DO NOT use wire or shoestring for a tourniquet band.

b. Placing the Improvised Tourniquet.

(1) Place the tourniquet around the limb, between the wound and the body trunk (or between the wound and the heart). Never place it directly over a wound, a fracture, or joint. Tourniquets, for maximum effectiveness, should be placed on the upper arm or above the knee on the thigh (Figure 2-32).

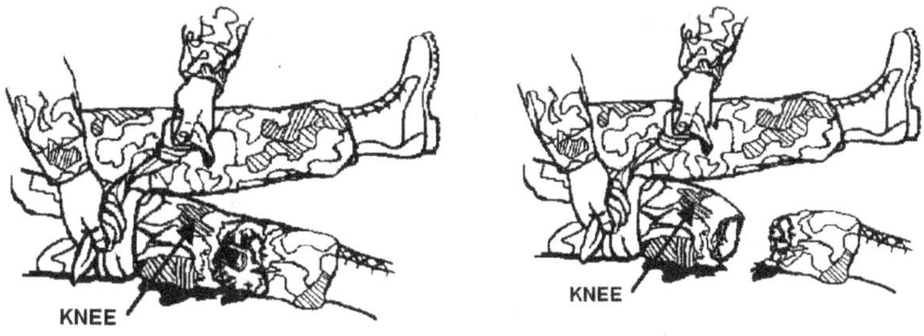

Figure 2-32. Tourniquet above knee.

(2) The tourniquet should be well-padded. If possible, place the tourniquet over the smoothed sleeve or trouser leg to prevent the skin from being pinched or twisted. If the tourniquet is long enough, wrap it around the limb several times, keeping the material as flat as possible. Damaging the skin may deprive the surgeon of skin required to cover an amputation. Protection of the skin also reduces pain.

c. Applying the Tourniquet.

(1) Tie a half-knot. (A half-knot is the same as the first part of tying a shoe lace.)

(2) Place a stick (or similar rigid object) on top of the half-knot (Figure 2-33).

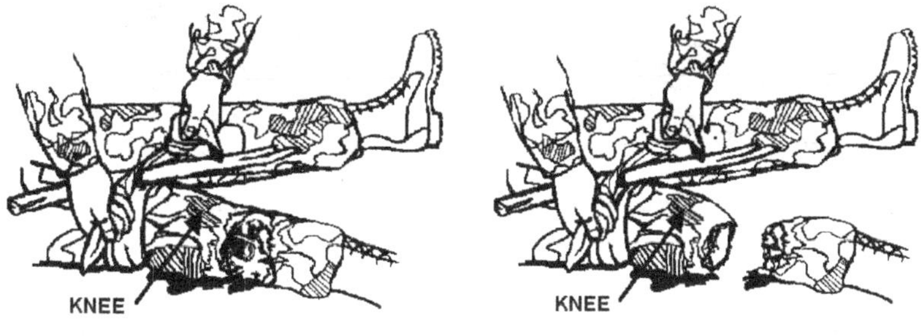

Figure 2-33. Rigid object on top of half-knot.

(3) Tie a full knot over the stick (Figure 2-34).

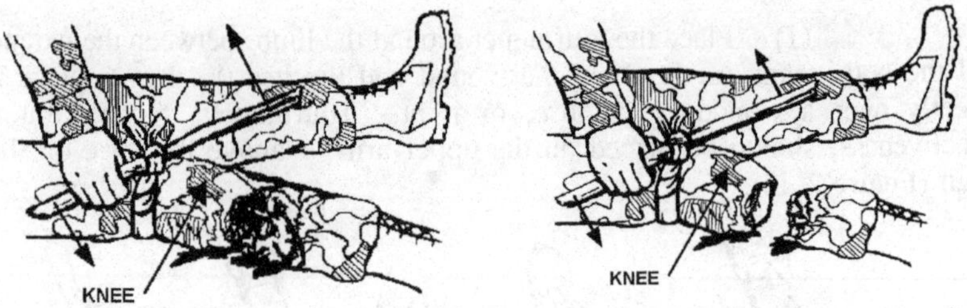

Figure 2-34. Full knot over rigid object.

(4) Twist the stick (Figure 2-35) until the tourniquet is tight around the limb and/or the bright red bleeding has stopped. In the case of amputation, dark oozing blood may continue for a short time. This is the blood trapped in the area between the wound and tourniquet.

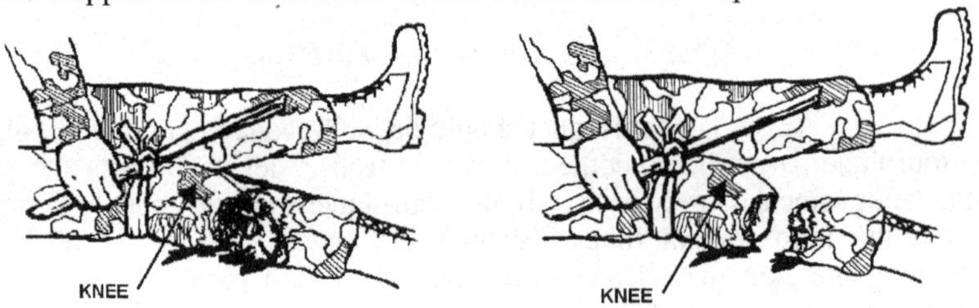

Figure 2-35. Stick twisted.

(5) Fasten the tourniquet to the limb by looping the free ends of the tourniquet over the ends of the stick. Then bring the ends around the limb to prevent the stick from loosening. Tie them together on the side of the limb (Figure 2-36).

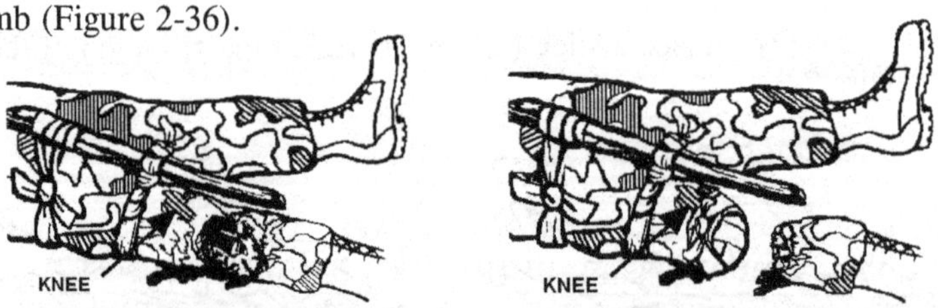

Figure 2-36. Tie free ends on side of limb.

NOTE

Other methods of securing the stick may be used as long as the stick does not unwind and no further injury results.

NOTE

If possible, save and transport any severed (amputated) limbs or body parts with (but out of sight of) the casualty.

(6) DO NOT cover the tourniquet—you should leave it in full view. If the limb is missing (total amputation), apply a dressing to the stump. All wounds should have a dressing to protect the wound from contamination.

(7) Mark the casualty's forehead with a "T" and the time to indicate a tourniquet has been applied. If necessary, use the casualty's blood to make this mark.

(8) Check and treat for shock.

(9) Seek medical aid.

CAUTION

Only appropriately skilled medical personnel may adjust or otherwise remove/release the tourniquet in the appropriate setting.

Section III. CHECK FOR SHOCK AND ADMINISTER FIRST AID MEASURES

2-21. General

The term *shock* has a variety of meanings. In medicine, it refers to a collapse of the body's cardiovascular system which includes an inadequate supply of blood to the body's tissues. Shock stuns and weakens the body. When the normal blood flow in the body is upset, death can result. Early recognition and proper first aid may save the casualty's life.

2-22. Causes and Effects

a. There are three basic mechanisms associated with shock. These are—

- The heart is damaged and fails to work as a pump.

- Blood loss (heavy bleeding) causes the volume of fluid within the vascular system to be insufficient.

- The blood vessels dilate (open wider) so that the blood within the system (even though it is a normal volume [the casualty is not bleeding or dehydrated]) is insufficient to provide adequate circulation within the body.

b. Shock may be the result of a number of conditions. These include—

- Dehydration.

- Allergic reaction to foods, drugs, insect stings, and snakebites.

- Significant loss of blood.

- Reaction to the sight of a wound, blood, or other traumatic scene.

- Traumatic injuries, such as—

  - Burns.

  - Gunshot or shrapnel wounds.

  - Crush injuries.

  - Blows to the body (which can cause broken bones or damage to internal organs).

  - Head injuries.

  - Penetrating wounds (such as from a knife, bayonet, or missile).

2-23. Signs and Symptoms of Shock

Examine the casualty to see if he has any of the following signs and symptoms:

- Sweaty but cool skin (clammy skin).

- Weak and rapid pulse.

- Paleness of skin (in dark-skinned individuals they may have a grayish look to their skin).

- Restlessness, nervousness.

- Thirst.

- Loss of blood (bleeding).

- Confusion (or loss of awareness).

- Faster-than-normal breathing rate.

- Blotchy or bluish skin (especially around the mouth and lips).

- Nausea and/or vomiting.

2-24. First Aid Measures for Shock

In the field, the first aid procedures administered for shock are identical to procedures that would be performed *to prevent shock*. When treating a casualty, assume that shock is present or will occur shortly. By waiting until actual signs and symptoms of shock are noticeable, the rescuer may jeopardize the casualty's life.

a. Position the Casualty. (DO NOT move the casualty or his limbs if suspected fractures have not been splinted. See Chapter 4 for details.)

(1) Move the casualty to cover, if cover is available and the situation permits.

(2) Lay the casualty on his back.

NOTE

A casualty in shock from a chest wound or one who is experiencing breathing difficulty, may breathe easier in a sitting position. If this is the case, allow him to sit upright, but monitor carefully in case his condition worsens.

(3) Elevate the casualty's feet higher than the level of his heart. Use a stable object (field pack or rolled up clothing) so that his feet will not slip off (Figure 2-37).

WARNING

DO NOT elevate legs if the casualty has an unsplinted broken leg, head injury, or abdominal injury.

Figure 2-37. Clothing loosened and feet elevated.

WARNING

Check casualty for leg fracture(s) and splint, if necessary, before elevating his feet. For a casualty with an abdominal wound, place his knees in an upright (flexed) position.

(4) Loosen clothing at the neck, waist, or wherever it may be binding.

CAUTION

DO NOT loosen or remove protective clothing in a chemical environment.

(5) Prevent chilling or overheating. The key is to maintain body temperature. In cold weather, place a blanket or other like item over him to keep him warm and under him to prevent chilling (Figure 2-38). However, if a tourniquet has been applied, leave it exposed (if possible). In hot weather, place the casualty in the shade and protect him from becoming chilled; however, avoid the excessive use of blankets or other coverings.

Figure 2-38. Body temperature maintained.

(6) Calm the casualty. Throughout the entire procedure of providing first aid for a casualty, the rescuer should reassure the casualty and keep him calm. This can be done by being authoritative (taking charge) and by showing self-confidence. Assure the casualty that you are there to help him.

(7) Seek medical aid.

b. Food and/or Drink. When providing first aid for shock, DO NOT give the casualty any food or drink. If you must leave the casualty or if he is unconscious, turn his head to the side to prevent him from choking if he vomits (Figure 2-39).

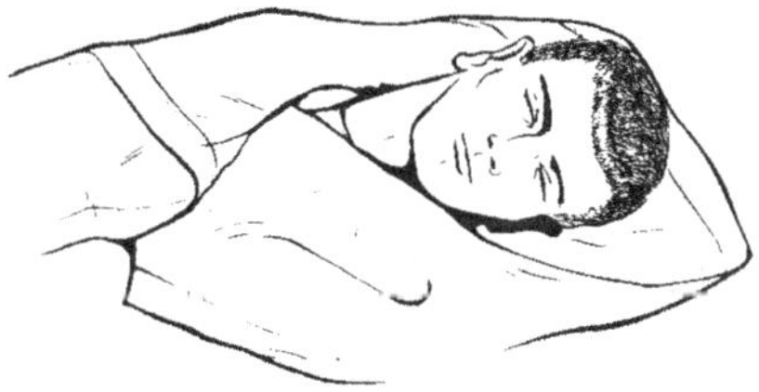

Figure 2-39. Casualty's head turned to side.

c. Evaluate Casualty. Continue to evaluate the casualty until medical personnel arrives or the casualty is transported to an MTF.

CHAPTER 3

FIRST AID FOR SPECIFIC INJURIES

3-1. General

Basic lifesaving steps are discussed in Chapters 1 and 2; they apply to first aid measures for all injuries. Some wounds and burns will require special precautions and procedures when applying these measures. This chapter discusses specific first aid procedures for wounds of the head, face, and neck; chest and stomach wounds; and burns. It also discusses the techniques for applying dressings and bandages to specific parts of the body.

3-2. Head, Neck, and Facial Injuries

a. Head Injuries.

(1) Head injuries range from minor abrasions or cuts on the scalp to severe brain injuries that may result in unconsciousness and sometimes death. Head injuries are classified as open or closed wounds. An open wound is one that is visible, has a break in the skin, and usually has evidence of bleeding. A closed wound may be visible (such as a depression in the skull) or the first aid provider may not be able to see any apparent injury (such as internal bleeding). Some head injuries result in unconsciousness; however, a service member may have a serious head wound and still be conscious. Casualties with head and neck injuries should be treated as though they also have a spinal injury. The casualty should not be moved until the head and neck is stabilized unless he is in immediate danger (such as close to a burning vehicle).

(2) Prompt first aid measures should be initiated for casualties with suspected head and neck injuries. The conscious casualty may be able to provide information on the extent of his injuries. However, as a result of the head injury, he may be confused and unable to provide accurate information. The signs and symptoms a first aid provider might observe are—

- Nausea and vomiting.
- Convulsions or twitches.
- Slurred speech.
- Confusion and loss of memory. (Does he know who he is? Does he know where he is? Does he know what day it is?)
- Recent unconsciousness.

- Dizziness.
- Drowsiness.
- Blurred vision, unequal pupils, or bruising (black eyes).
- Paralysis (partial or full).
- Complaint of headache.
- Bleeding or other fluid discharge from the scalp, nose, or ears.
- Deformity of the head (depression or swelling).
- Staggering while walking.

b. Neck Injuries. Neck injuries may result in heavy bleeding. Apply pressure above and below the injury, *but do not interfere with the breathing process*, and attempt to control the bleeding. Apply a dressing. Always evaluate the casualty for a possible neck fracture/spinal cord injury; if suspected, seek medical treatment immediately.

NOTE

Establish and maintain the airway in cases of facial or neck injuries. If a neck fracture or spinal cord injury is suspected, immobilize the injury and, if necessary, perform basic life support measures.

c. Facial Injuries. Soft tissue injuries of the face and scalp are common. Abrasions (scrapes) of the skin cause no serious problems. Contusions (injury without a break in the skin) usually cause swelling. A contusion of the scalp looks and feels like a lump. Laceration (cut) and avulsion (torn away tissue) injuries are also common. Avulsions are frequently caused when a sharp blow separates the scalp from the skull beneath it. Because the face and scalp are richly supplied with blood vessels (arteries and veins), wounds of these areas usually bleed heavily.

3-3. General First Aid Measures

a. General Considerations. The casualty with a head injury (or suspected head injury) should be continually monitored for the development of conditions that *may require* basic lifesaving measures. After initiating first aid measures, request medical assistance and evacuation. If dedicated medical evacuation assets are not available, transport the casualty to an MTF as soon as the situation permits. The first aid provider should not attempt to remove a protruding object from the head or give the casualty anything to eat or drink. Further, the first aid provider should be prepared to—

- Clear the airway.
- Control bleeding (external).
- Administer first aid measures for shock.
- Keep the casualty warm.
- Protect the wound.

b. Unconscious Casualty. An unconscious casualty does not have control of all of his body's functions and may choke on his tongue, blood, vomitus, or other substances. (Refer to Figure 2-39.)

(1) *Breathing.* The brain requires a constant supply of oxygen. A bluish (or in an individual with dark skin—grayish) color of skin around the lips and nail beds indicates that the casualty is not receiving enough oxygen. Immediate action must be taken to clear the airway, to position the casualty on his side, or to initiate rescue breathing.

(2) *Bleeding.* Bleeding from a head injury usually comes from blood vessels within the scalp. Bleeding can also develop inside the skull or within the brain. In most instances visible bleeding from the head can be controlled by application of the field first aid dressing.

CAUTION

DO NOT attempt to put unnecessary pressure on the wound or attempt to push any brain matter back into the head (skull). **DO NOT** apply a pressure dressing.

c. Concussion. If an individual receives a heavy blow to the head or face, he may suffer a brain concussion (an injury to the brain that involves a temporary loss of some or all of the brain's ability to function). For example, the casualty may not breathe properly for a short period of time, or he may become confused and stagger when he attempts to walk. Symptoms of a concussion may only last for a short period of time. However, if a casualty is suspected of having suffered a concussion, he should be transported to an MTF as soon as conditions permit.

d. Convulsions. Convulsions (seizures/involuntary jerking) may occur even after a mild head injury. When a casualty is convulsing, protect him from hurting himself. Take the following measures:

(1) Ease him to the ground if he is standing or sitting.

(2) Support his head and neck.

(3) Maintain his airway.

(4) Protect him from further injury (such as hitting close-by objects).

NOTE

DO NOT forcefully hold the arms and legs if they are jerking because this can lead to broken bones. **DO NOT** force anything between the casualty's teeth—especially if they are tightly clenched because this may obstruct the casualty's airway. Maintain the casualty's airway if necessary.

e. Brain Damage. In *severe* head injuries where brain tissue is protruding, *leave the wound alone*; carefully place a loose moistened dressing (moistened with sterile normal saline if available) and also a first aid dressing over the tissue to protect it from further contamination. ***DO NOT** remove or disturb any foreign matter that may be in the wound.* Position the casualty so that his head is higher than his body. Keep him warm and *seek medical assistance immediately.*

NOTE

If there is an object extending from the wound, **DO NOT** remove the object. Improvise bulky dressings from the cleanest material available and place this material around the protruding object for support, then apply the field dressing.

3-4. Chest Wounds

Blunt trauma, bullet or missile wounds, stab wounds, or falls may cause chest injuries. These injuries can be serious and may cause death quickly if first aid is not administered in a timely manner. A casualty with a chest injury may complain of pain in the chest or shoulder area; he may have difficulty breathing. His chest may not rise normally when he breathes. The injury may cause the casualty to cough up blood and to have a rapid or a weak heartbeat. A casualty with an open chest wound has a punctured chest wall. The sucking sound heard when he breathes is caused by air leaking into his chest cavity. This particular type of wound is dangerous and will collapse the injured lung (Figure 3-1). Breathing becomes difficult for the casualty because the wound is open. The service members life may depend upon how quickly you apply an occlusive dressing over the wound (refer to paragraph 3-5).

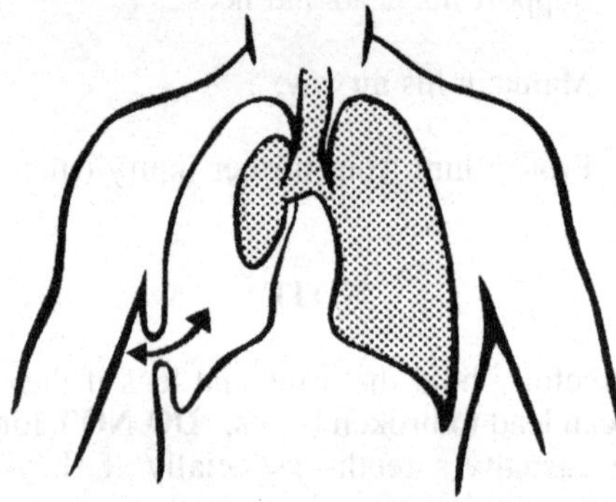

Figure 3-1. Collapsed lung.

3-5. First Aid for Chest Wounds

a. Evaluate the Casualty. Be prepared to perform first aid measures. These measures may include clearing the airway, rescue breathing, treatment for shock, and/or bleeding control.

b. Expose the Wound. If appropriate, cut or remove the casualty's clothing to expose the wound. Remember, **DO NOT** remove clothing that is stuck to the wound because additional injury may result. **DO NOT** attempt to clean the wound.

NOTE

Examine the casualty to see if there is an entry and exit wound. If there are two wounds (entry, exit), perform the same procedure for both wounds. Treat the more serious (heavier bleeding, larger) wound first. It may be necessary to improvise a dressing for the second wound by using strips of cloth, such as a torn T-shirt, or whatever material is available. Also, listen for sucking sounds to determine if the chest wall is punctured.

CAUTION

If there is an object impaled in the wound, **DO NOT** remove it. Apply a dressing around the object and use additional improvised bulky materials/dressings (use the cleanest materials available) to build up the area around the object. Apply a supporting bandage over the bulky materials to hold them in place.

CAUTION

DO NOT REMOVE protective clothing in a chemical environment. Apply dressings *over* the protective clothing.

c. *Open the Casualty's Field Dressing Plastic Wrapper.* In cases where there is a sucking chest wound, the plastic wrapper is used with the field dressing to create an occlusive dressing. If a plastic wrapper is not available, or if an additional wound needs to be treated; cellophane, foil, the casualty's poncho, or similar material may be used. The covering should be wide enough to extend 2 inches or more beyond the edges of the wound in all directions.

(1) Tear open one end of the casualty's plastic wrapper covering the field dressing. Be careful not to destroy the wrapper and **DO NOT** touch the inside of the wrapper.

(2) Remove the inner packet (field dressing).

(3) Complete tearing open the empty plastic wrapper using as much of the wrapper as possible to create a flat surface.

d. *Place the Wrapper Over the Wound.* Place the inside surface of the plastic wrapper directly over the wound *when the casualty exhales* and hold it in place (Figure 3-2). The casualty may hold the plastic wrapper in place if he is able.

Figure 3-2. Open chest wound sealed with an occlusive dressing.

e. Apply the Dressing to the Wound.

(1) Use your free hand and shake open the field dressing (Figure 3-3).

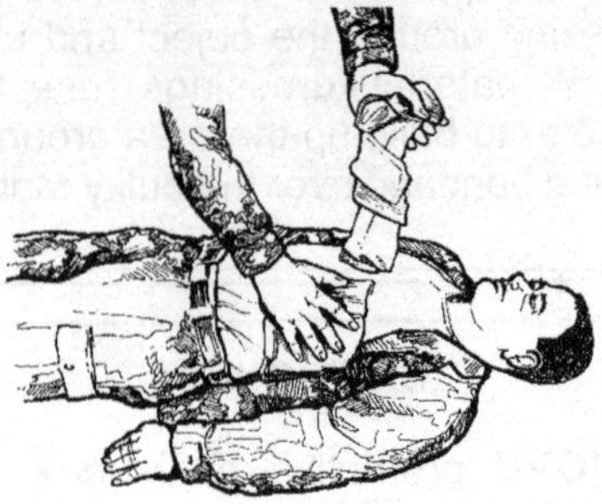

Figure 3-3. Shaking open the field dressing.

(2) Place the white side of the dressing on the plastic wrapper covering the wound (Figure 3-4).

Figure 3-4. Field dressing placed on plastic wrapper.

NOTE

Use the casualty's field dressing, not your own.

(3) Have the casualty breathe normally.

(4) While maintaining pressure on the dressing, grasp one tail of the field dressing with the other hand and wrap it around the casualty's back. If tape is available, tape three sides of the plastic wrapper to the chest wall to provide occlusive type dressing. Leave one side untapped to provide emergency escape for air that may build up in the chest. If tape is not available, secure wrapper on three sides with field dressing leaving the fourth side as a flap.

(5) Wrap the other tail in the opposite direction, bringing both tails over the dressing (Figure 3-5).

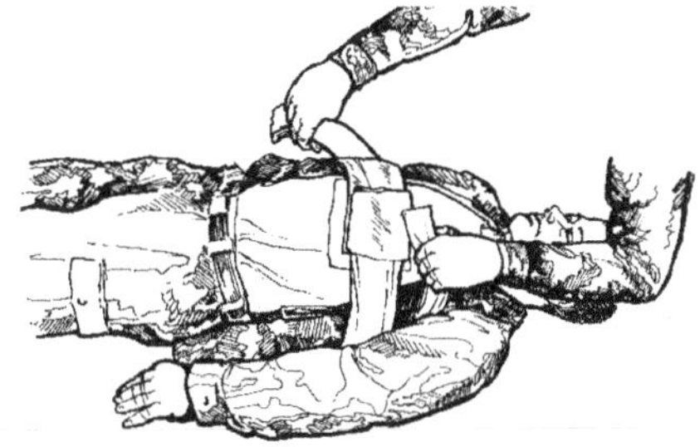

Figure 3-5. Tails of field dressing wrapped around casualty in opposite direction.

(6) Tie the tails into a square knot in the center of the dressing *after* the casualty exhales and *before* he inhales. This will aid in maintaining pressure on the bandage after it has been tied (Figure 3-6). Tie the dressing firmly enough to secure the dressing without interfering with the casualty's breathing.

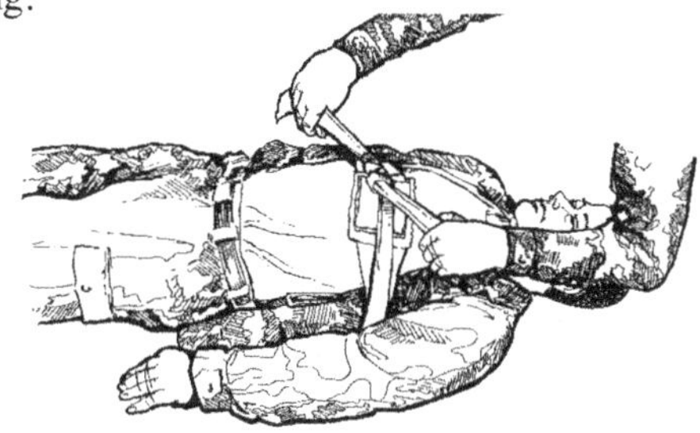

Figure 3-6. Tails of dressing tied into square knot over center of dressing.

NOTE

When practical, apply direct manual pressure over the dressing for 5 to 10 minutes to help control the bleeding.

f. Position the Casualty. Position the casualty on his injured side or in a sitting position, whichever makes breathing easier (Figure 3-7).

Figure 3-7. Casualty positioned (lying) on injured side.

g. Seek Medical Assistance. Contact medical personnel.

WARNING

If an occlusive dressing has been improperly placed, air may enter the chest cavity with no means of escape. This causes a life-threatening condition called *tension pneumothorax*. If the casualty's condition (for example, difficulty breathing, shortness of breath, restlessness, or blueness/grayness of the skin) worsens after placing the dressing, quickly lift or remove, and then replace the occlusive dressing.

3-6. Abdominal Wounds

The most serious abdominal wound is one in which an object penetrates the abdominal wall and pierces internal organs or large blood vessels. In these instances, bleeding may be severe and death can occur rapidly.

3-7. First Aid for Abdominal Wounds

a. Evaluate the Casualty. Be prepared to perform basic first aid measures. Always check for both entry and exit wounds. If there are two wounds (entry and exit), treat the wound that appears more serious first (for example, the heavier bleeding, protruding organs, larger wound, and so forth). It may be necessary to improvise dressings for the second wound by using strips of cloth, a T-shirt, or the cleanest material available.

b. Position the Casualty. Place and maintain the casualty on his back with his knees in an upright (flexed) position (Figure 3-8). The knees-up position helps relieve pain, assists in the treatment of shock, prevents further exposure of the bowel (intestines) or abdominal organs, and helps relieve abdominal pressure by allowing the abdominal muscles to relax.

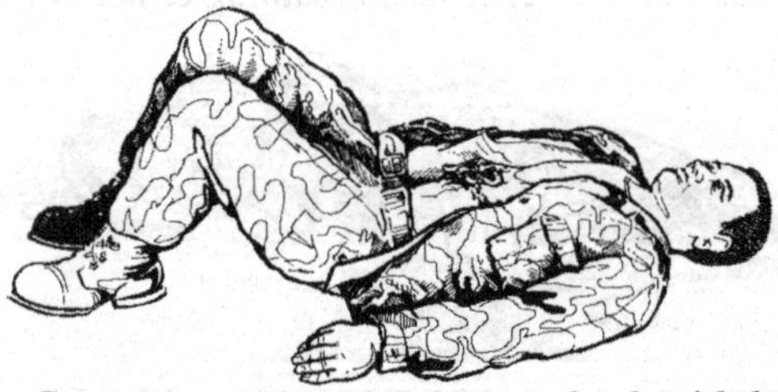

Figure 3-8. Casualty positioned (lying) on back with knees (flexed) up.

c. Expose the Wound.

(1) Remove the casualty's loose clothing to expose the wound. However, **DO NOT** attempt to remove clothing that is stuck to the wound; removing it may cause further injury.

CAUTION

DO NOT REMOVE protective clothing in a chemical environment. Apply dressings *over* the protective clothing.

(2) Gently pick up any organs that may be on the ground. Do this with a clean, dry dressing or with the cleanest available material. Place the organs on top of the casualty's abdomen (Figure 3-9).

Figure 3-9. Protruding organs placed near wound.

NOTE

DO NOT probe, clean, or try to remove any foreign object from the abdomen. **DO NOT** touch with bare hands any exposed organs. **DO NOT** push organs back inside the body.

d. Apply the Field Dressing. Use the casualty's field dressing, not your own. If the field dressing is not large enough to cover the entire wound, the plastic wrapper from the dressing may be used to cover the wound first (placing the field dressing on top). Open the plastic wrapper carefully without touching the inner surface. If necessary, other improvised dressings may be made from clothing, blankets, or the cleanest materials available.

WARNING

If there is an object extending from the wound, DO NOT remove it. Place as much of the wrapper over the wound as possible without dislodging or moving the object. DO NOT place the wrapper over the object.

(1) Grasp the tails in both hands.

(2) Hold the dressing with the white side down directly over the wound. **DO NOT** touch the white (sterile) side of the dressing or allow anything except the wound to come in contact with it.

(3) Pull the dressing open and place it directly over the wound (Figure 3-10). If the casualty is able, he may hold the dressing in place.

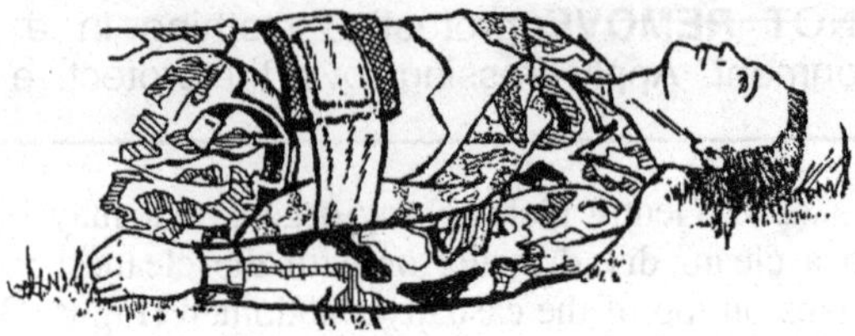

Figure 3-10. Dressing placed directly over the wound.

(4) Hold the dressing in place with one hand and use the other hand to wrap one of the tails around the body.

(5) Wrap the other tail in the opposite direction until the dressing is completely covered. Leave enough of the tail for a knot.

(6) Loosely tie the tails with a square knot at the casualty's side (Figure 3-11).

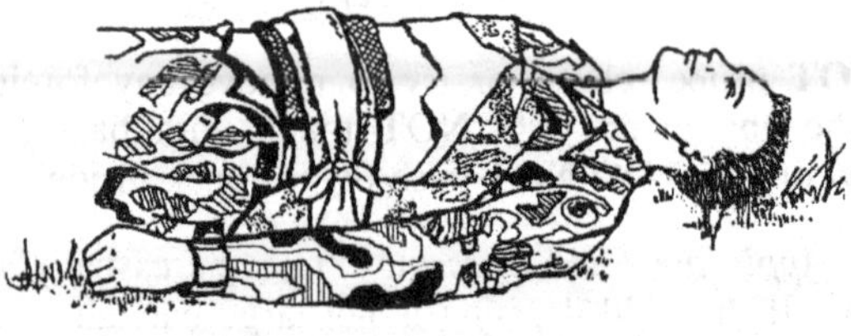

Figure 3-11. Dressing applied and tails tied with a square knot.

WARNING

When the dressing is applied, DO NOT put pressure on the wound or exposed internal parts, because pressure could cause further injury (vomiting, ruptured intestines, and so forth). Therefore, tie the dressing ties (tails) loosely at casualty's side, not directly over the dressing.

(7) Tie the dressing firmly enough to prevent slipping without applying pressure to the wound site (Figure 3-12).

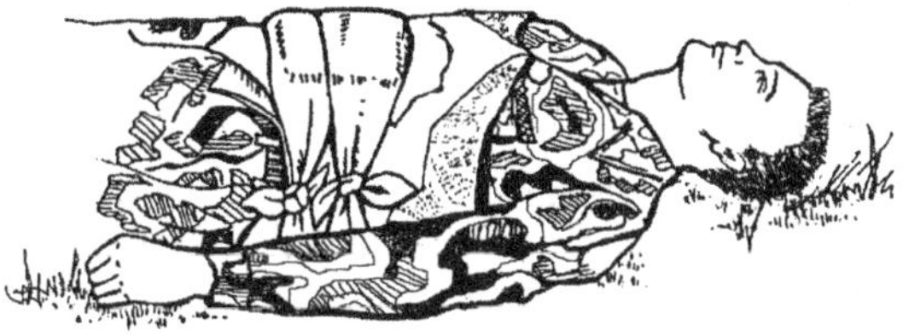

Figure 3-12. Field dressing covered with improvised material and loosely tied.

Field dressings can be covered with improvised reinforcement material (cravats, strips of torn T-shirt, or other cloth) for additional support and protection. Tie improvised bandage on the opposite side of the dressing ties firmly enough to prevent slipping but without applying additional pressure to the wound.

> **CAUTION**
>
> **DO NOT** give casualties with abdominal wounds food or water (moistening the lips is allowed).

e. Seek Medical Assistance. Notify medical personnel.

3-8. Burn Injuries

Burns often cause extreme pain, scarring, or even death. Before administering first aid, you must be able to recognize the type of burn. There are four types of burns:

- Thermal burns caused by fire, hot objects, hot liquids, and gases; or by nuclear blast or fireball.
- Electrical burns caused by electrical wires, current, or lightning.
- Chemical burns caused by contact with wet or dry chemicals or white phosphorus (WP)—from marking rounds and grenades.
- Laser burns (eye [ocular] injury).

3-9. First Aid for Burns

a. Eliminate the Source of the Burn. The source of the burn must be eliminated before any evaluation of the casualty can occur and first aid administered.

(1) Quickly remove the casualty from danger and cover the *thermal burn* with any large nonsynthetic material, such as a field jacket. If the casualty's clothing is still on fire, roll the casualty on the ground to smother (put out) the flames (Figure 3-13).

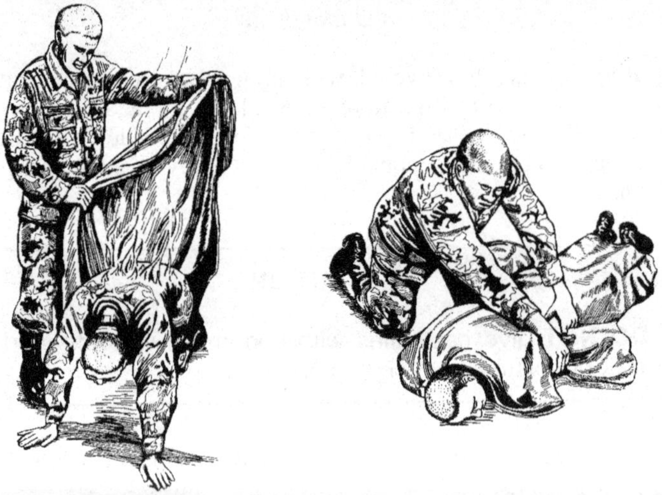

Figure 3-13. Casualty covered and rolled on ground.

CAUTION

Synthetic materials, such as nylon, may melt and cause further injury.

(2) Remove the *electrical burn* casualty from the electrical source by turning off the electrical current. **DO NOT** attempt to turn off the electricity if the source is not close by. Speed is critical, so **DO NOT** waste unnecessary time. If the electricity cannot be turned off, wrap any *nonconductive* material (*dry* rope, clothing, wood, and so forth) around the casualty's back and shoulders and drag the casualty away from the electrical source (Figure 3-14). **DO NOT** make body-to-body contact with the casualty or touch any wires because you could also become an electrical burn casualty.

Figure 3-14. Casualty removed from electrical source (using nonconductive material).

WARNING

High voltage electrical burns may cause temporary unconsciousness, difficulties in breathing, or difficulties with the heart (heartbeat).

(3) Remove the *chemical* from the *burned casualty*. Remove *liquid* chemicals by flushing with as much water as possible. Remove *dry* chemicals by brushing off loose particles (**DO NOT** use the bare surface of your hand because you could become a chemical burn casualty) and then flush with large amounts of water, if available. If large amounts of water are not available, then **NO** water should be applied because small amounts of water applied to a dry chemical burn may cause a chemical reaction. When WP strikes the skin, smother with a wet cloth or mud. Keep WP covered with a wet material to exclude air; this should help prevent the particles from burning.

(4) Remove the *laser burn* casualty from the source. When removing the casualty from the laser beam source, be careful not to enter the

beam or you may become a casualty. Never look directly at the beam source and if possible, wear appropriate eye protection.

NOTE

After the casualty is removed from the source of the burn, he should be evaluated for conditions requiring basic first aid measures.

b. Expose the Burn. Cut and gently lift away any clothing covering the burned area, without pulling clothing over the burns. Leave in place any clothing that is stuck to the burn. If the casualty's hands or wrists have been burned, remove jewelry if possible without causing further injury (rings, watches, and so forth) and place in his pockets. This prevents the necessity to cut off jewelry since swelling usually occurs as a result of a burn.

CAUTION

DO NOT lift or cut away clothing if in a chemical environment. Apply the dressing directly over the casualty's protective clothing. **DO NOT** attempt to decontaminate skin where blisters have formed.

c. Apply a Field Dressing to the Burn.

(1) Grasp the tails of the casualty's dressing in both hands.

(2) Hold the dressing directly over the wound with the white side down, pull the dressing open, and place it directly over the wound. **DO NOT** touch the white (sterile) side of the dressing or allow anything except the wound to come in contact with it. If the casualty is able, he may hold the dressing in place.

(3) Hold the dressing in place with one hand and use the other hand to wrap one of the tails around the limbs or the body.

(4) Wrap the other tail in the opposite direction until the dressing is completely covered.

(5) Tie the tails into a square knot over the outer edge of the dressing. The dressing should be applied lightly over the burn. Ensure that dressing is applied firmly enough to prevent it from slipping.

NOTE

Use the cleanest improvised dressing material available if a field dressing is not available or if it is not large enough for the entire wound.

d. *Take the Following Precautions*:

- **DO NOT** place the dressing over the face or genital area.

- **DO NOT** break the blisters.

- **DO NOT** apply grease or ointments to the burns.

- For electrical burns, check for both an entry and exit burn from the passage of electricity through the body. Exit burns may appear on any area of the body despite location of entry burn.

- For burns caused by wet or dry chemicals, flush the burns with large amounts of water and cover with a dry dressing.

- For burns caused by WP, flush the area with water, then cover with a wet material, dressing, or mud to exclude the air and keep the WP particles from burning.

- For laser burns, apply a field dressing.

- If the casualty is conscious and not nauseated, give him small amounts of water.

e. *Seek Medical Assistance*. Notify medical personnel.

3-10. Dressings and Bandages

a. *Head Wounds.*

(1) *Position the casualty.*

WARNING

DO NOT move the casualty if you suspect he has sustained a neck, spine, or head injury (which produces any signs or symptoms other than minor bleeding).

• If the casualty has a minor (superficial) scalp wound and is conscious:

• Have the casualty sit up (unless other injuries prohibit or he is unable to).

• If the casualty is lying down and is not accumulating fluids or drainage in his throat, elevate his head slightly.

• If the casualty is bleeding from or into his mouth or throat, turn his head to the side or position him on his side so that the airway will be clear. Avoid putting pressure on the wound and place him on his uninjured side (Figure 3-15).

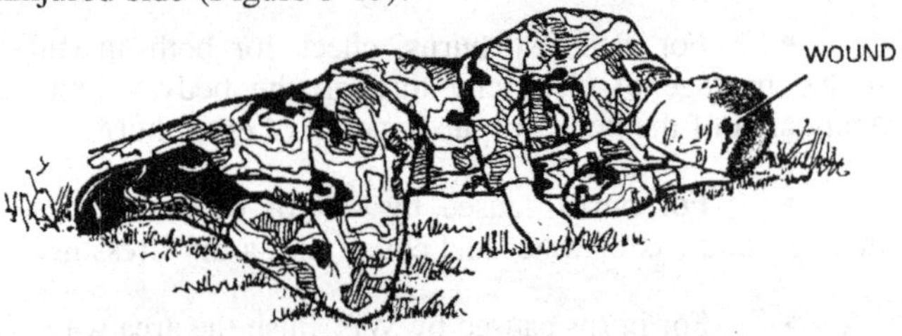

Figure 3-15. Casualty lying on side opposite injury.

• If the casualty is unconscious or has a severe head injury, then suspect and treat him as having a potential neck or spinal injury, *immobilize and **DO NOT** move the casualty.*

NOTE

If the casualty is choking or vomiting or is bleeding from or into his mouth (thus compromising his airway), position him on his uninjured side to allow for drainage and to help keep his airway clear.

WARNING

If it is necessary to turn a casualty with a suspected neck/spine injury; roll the casualty gently onto his side, keeping the head, neck, and body aligned while providing support for the head and neck. DO NOT roll the casualty by yourself but seek assistance. *Move him only if absolutely necessary*, otherwise keep the casualty immobilized to prevent further damage to the neck/spine.

(2) *Expose the wound.* Remove the casualty's helmet (if necessary). In a nuclear, biological, and chemical (NBC) environment, the first aid provider must leave the casualty as much protection (such as protective mask, mission-oriented protective posture [MOPP] overgarments) as possible. What items of protective equipment can be removed is dependent upon the casualty's injuries (where on the body and what type), the MOPP level, integrity of protective equipment (such as tears in the garment or mask seal), availability of chemical protective shelters, and the tactical situation.

WARNING

DO NOT attempt to clean the wound or remove a protruding object.

NOTE

Always use the casualty's field dressing, not your own.

(3) *Apply a dressing to a wound of the forehead or back of head.* To apply a dressing to a wound of the forehead or back of the head—

(*a*) Remove the dressing from the wrapper.

(*b*) Grasp the tails of the dressing in both hands.

(*c*) Hold the dressing (white side down) directly over the wound. **DO NOT** touch the white (sterile) side of the dressing or allow anything except the wound to come in contact with it.

(*d*) Place it directly over the wound.

(*e*) Hold it in place with one hand. If the casualty is able, he may assist.

(*f*) Wrap the first tail horizontally around the head; ensure the tail covers the dressing (Figure 3-16).

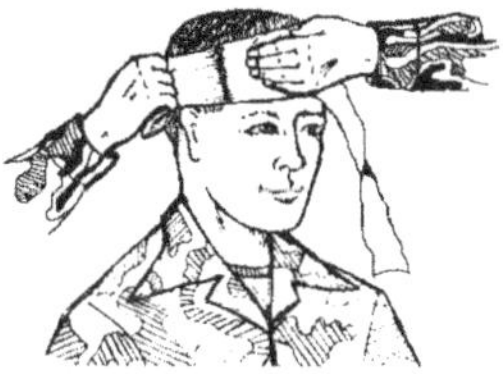

Figure 3-16. First tail of dressing wrapped horizontally around head.

(*g*) Hold the first tail in place and wrap the second tail in the opposite direction, covering the dressing (Figure 3-17).

Figure 3-17. Second tail wrapped in opposite direction.

(*h*) Tie a square knot and secure the tails at the side of the head, making sure they **DO NOT** cover the eyes or ears (Figure 3-18).

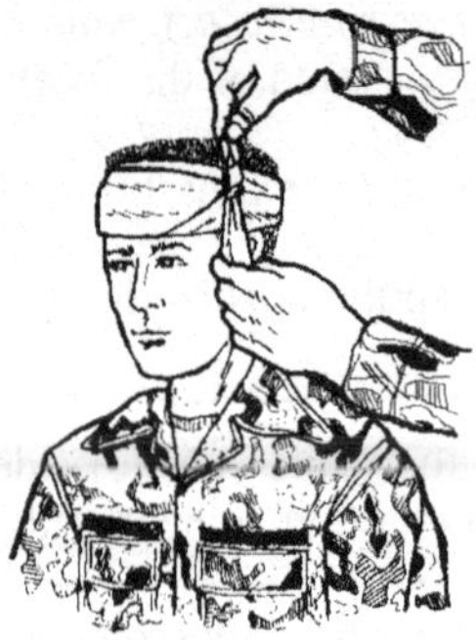

Figure 3-18. Tails tied in square knot at side of head.

(4) *Apply a dressing to a wound on top of the head.* To apply a dressing to a wound on top of the head—

(*a*) Remove the dressing from the wrapper.

(*b*) Grasp the tails of the dressing in both hands.

(*c*) Hold it (white side down) directly over the wound. **DO NOT** touch the white (sterile) side of the dressing or allow anything except the wound to come in contact with it.

(*d*) Place it over the wound (Figure 3-19).

Figure 3-19. Dressing placed over wound.

(*e*) Hold it in place with one hand. If the casualty is able, he may assist.

(*f*) Wrap one tail down under the chin (Figure 3-20), up in front of the ear, over the dressing, and in front of the other ear.

Figure 3-20. One tail of dressing wrapped under chin.

WARNING

Ensure the tails remain wide and close to the front of the chin to avoid choking the casualty.

(*g*) Wrap the remaining tail under the chin in the opposite direction and up the side of the face to meet the first tail (Figure 3-21).

Figure 3-21. Remaining tail wrapped under chin in opposite direction.

(*h*) Cross the tails (Figure 3-22), bringing one around the forehead (above the eyebrows) and the other around the back of the head (at the base of the skull) to a point just above and in front of the opposite ear, and tie them using a square knot (Figure 3-23).

Figure 3-22. Tails of dressing crossed with one around forehead.

Figure 3-23. Tails tied in square knot (in front of and above ear).

(5) *Apply a triangular bandage to the head.* To apply a triangular bandage to the head—

(*a*) Turn the base (longest side) of the bandage up and center its base on the center of the forehead, letting the point (apex) fall on the back of the neck (Figure 3-24A).

(*b*) Take the ends behind the head and cross the ends over the apex.

(*c*) Take them over the forehead and tie them (Figure 3-24B).

(*d*) Tuck the apex behind the crossed part of the bandage or secure it with a safety pin, if available (Figure 3-24C).

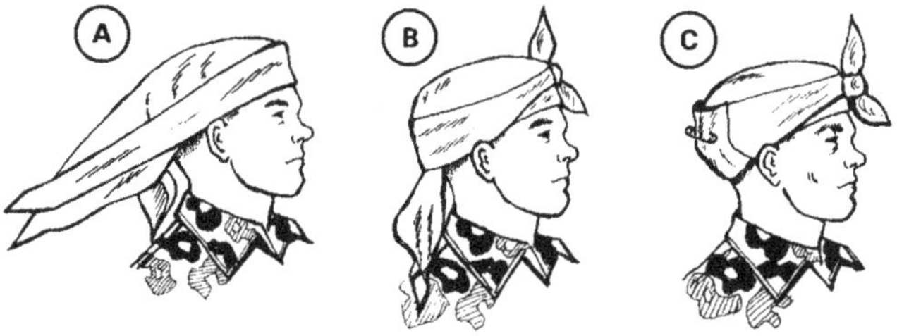

Figure 3-24. Triangular bandage applied to head (Illustrated A—C)

(6) *Apply a cravat bandage to the head.* To apply a cravat bandage to the head—

(*a*) Place the middle of the bandage over the dressing (Figure 3-25A).

(*b*) Cross the two ends of the bandage in opposite directions completely around the head (Figure 3-25B).

(*c*) Tie the ends over the dressing (Figure 3-25C).

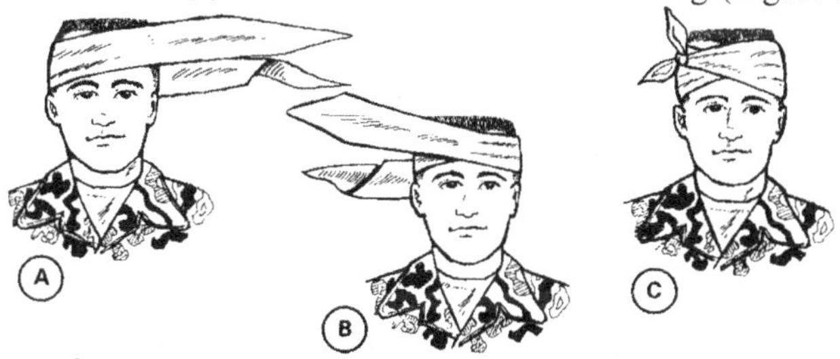

Figure 3-25. Cravat bandage applied to head (Illustrated A—C).

b. *Eye Injuries*. The eye is a vital sensory organ, and blindness is a severe physical handicap. Timely first aid of the eye may relieve pain and may also help to prevent shock, permanent eye injury, and possible loss of vision. Because the eye is very sensitive, any injury can be easily aggravated if it is improperly handled. Injuries of the eye may be quite severe. Cuts of the eyelids can appear to be very serious, but if the eyeball is not involved, a person's vision usually will not be damaged. However, lacerations (cuts) of the eyeball can cause permanent damage or loss of sight.

(1) *Lacerated/torn eyelids*. Lacerated eyelids may bleed heavily, but bleeding usually stops quickly. Cover the injured eye with a sterile dressing. **DO NOT** put pressure on the wound because you may injure the eyeball. Handle torn eyelids very carefully to prevent further injury. Place any detached pieces of the eyelid on a clean bandage or dressing and immediately send them with the casualty to the medical facility.

(2) *Lacerated eyeball (injury to the globe)*. Lacerations or cuts to the eyeball may cause serious and permanent eye damage. Cover the injury with a loose sterile dressing. **DO NOT** put pressure on the eyeball because additional damage may occur. An important point to remember is that when one eyeball is injured, you should immobilize both eyes. This is done by applying a bandage to both eyes. Because the eyes move together, covering both will lessen the chances of further damage to the injured eye. (However, in hazardous surroundings, leave uninjured eye uncovered to enable casualty to see.)

CAUTION

DO NOT apply pressure when there is a possible laceration of the eyeball. The eyeball contains fluid. Pressure applied over the eye will force the fluid out, resulting in permanent injury. **APPLY PROTECTIVE DRESSING WITHOUT ADDED PRESSURE**.

(3) *Extruded eyeballs*. Service members may encounter casualties with severe eye injuries that include an extruded eyeball (eyeball out-of-socket). In such instances you should gently cover the extruded eye with a loose moistened dressing and also cover the unaffected eye. **DO NOT** bind or exert pressure on the injured eye while applying the dressing. Keep the casualty quiet, place him on his back, treat for shock, and evacuate him immediately.

(4) *Burns of the eyes*. Chemical burns, thermal (heat) burns, and light burns can affect the eyes.

(*a*) *Chemical burns*. Injuries from chemical burns require immediate first aid. Mainly acids or alkalies cause chemical burns. The first aid measures consist of flushing the eyes immediately with large amounts of water for at least 5 to 20 minutes, or as long as necessary to flush out the chemical and, once flushed, bandaging the eyes. If the burn is an acid burn, you should flush the eye for at least 5 to 10 minutes. If the burn is an alkali burn, you should flush the eye for at least 20 minutes. After the eye has been flushed evacuate the casualty immediately.

(*b*) *Thermal burns*. When an individual suffers burns of the face from a fire, the eyes will close quickly due to extreme heat. This reaction is a natural reflex to protect the eyeballs; however, the eyelids remain exposed and are frequently burned. If a casualty receives burns of the eyelids or face—

- **DO NOT** apply a dressing.
- **DO NOT** touch.
- **SEEK** medical assistance immediately.

(*c*) *Light burns*. Exposure to intense light can burn an individual. Infrared rays, eclipse light (if the casualty has looked directly at the sun), or laser burns cause injuries of the exposed eyeball. Ultraviolet rays from arc welding can cause a superficial burn to the surface of the eye. These injuries are generally not painful but may cause permanent damage to the eyes. Immediate first aid is usually not required. Loosely bandaging the eyes may make the casualty more comfortable and protect his eyes from further injury caused by exposure to other bright lights or sunlight.

CAUTION

With impaled objects or significant sized foreign bodies, both eyes are usually bandaged to help secure the foreign body in the injured eye. In a battlefield environment, leave the uninjured eye uncovered so that the casualty can see.

c. *Side-of-Head or Cheek Wound*. Facial injuries to the side of the head or the cheek may bleed profusely (Figure 3-26). Prompt action is necessary to ensure that the airway remains open and also to control the bleeding. It may be necessary to apply a dressing. To apply a dressing—

(1) Remove the dressing from its wrapper.

(2) Grasp the tails in both hands.

(3) Hold the dressing directly over the wound with the white side down and place it directly on the wound (Figure 3-27A). **DO NOT** touch the white (sterile) side of the dressing or allow anything except the wound to come in contact with it.

(4) Hold the dressing in place with one hand (the casualty may assist if able). Wrap the top tail over the top of the head and bring it down in front of the ear (on the side opposite the wound), under the chin (Figure 3-27B) and up over the dressing to a point just above the ear (on the wound side).

Figure 3-26. Side of head or cheek wound.

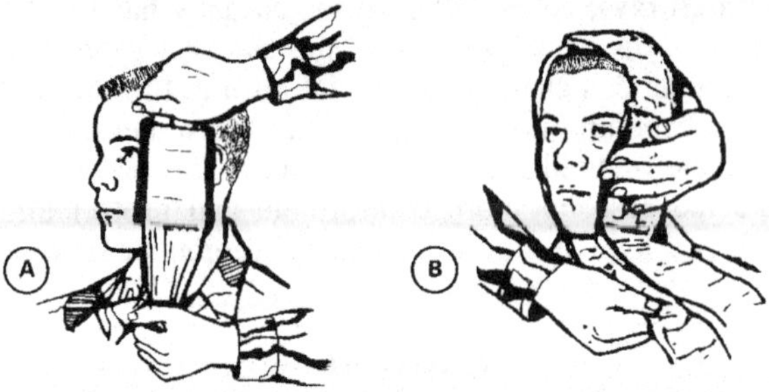

Figure 3-27. Dressing placed directly on wound. Top tail wrapped over top of head, down in front of ear, and under chin (Illustrated A—B).

NOTE

When possible, avoid covering the casualty's ear with the dressing, as this will decrease his ability to hear.

(5) Bring the second tail under the chin, up in front of the ear (on the side opposite the wound), and over the head to meet the other tail (on the wounded side) (Figure 3-28).

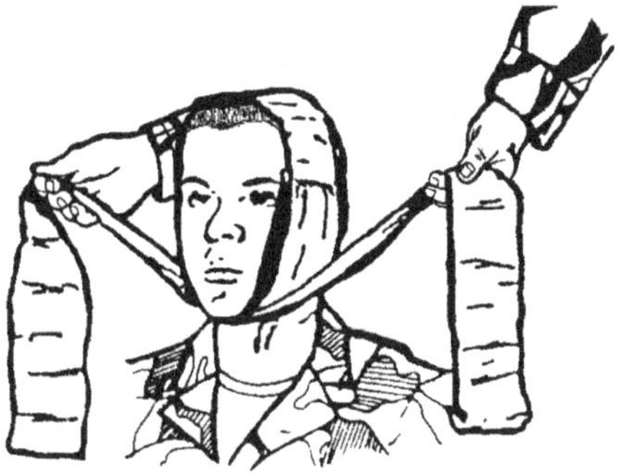

Figure 3-28. Bringing second tail under the chin.

(6) Cross the two tails (on the wound side) (Figure 3-29) and bring one end across the forehead (above the eyebrows) to a point just in front of the opposite ear (on the uninjured side).

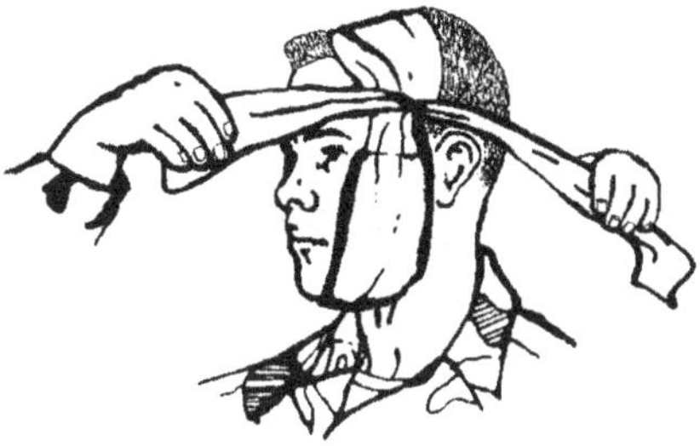

Figure 3-29. Crossing the tails on the side of the wound.

(7) Wrap the other tail around the back of the head (at the base of the skull), and tie the two ends just in front of the ear on the uninjured side with a square knot (Figure 3-30).

Figure 3-30. Tying the tails of the dressing in a square knot.

d. Ear Injuries. Lacerated (cut) or avulsed (torn) ear tissue may not, in itself, be a serious injury. Bleeding, or the drainage of fluids from the ear canal, however, may be a sign of a head injury, such as a skull fracture. **DO NOT** attempt to stop the flow from the inner ear canal nor put anything into the ear canal to block it. Instead, you should cover the ear lightly with a dressing. For minor cuts or wounds to the external ear, apply a cravat bandage as follows:

(1) Place the middle of the bandage over the ear (Figure 3-31A).

(2) Cross the ends, wrap them in opposite directions around the head, and tie them (Figures 3-31B and 3-31C).

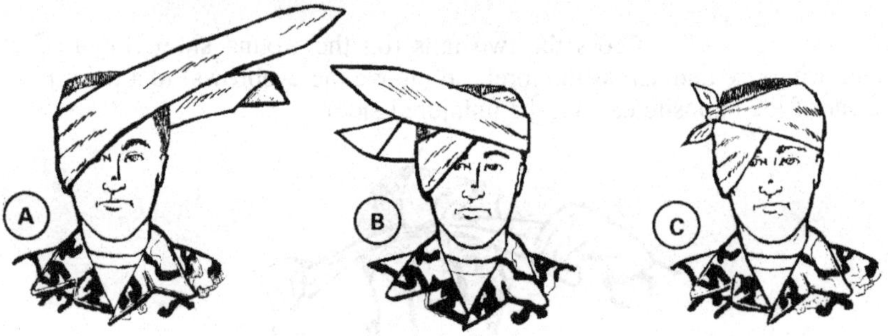

Figure 3-31. Applying cravat bandage to ear (Illustrated A—C).

(3) If possible, place some dressing material between the back of the ear and the side of the head to avoid crushing the ear against the head with the bandage.

e. Nose Injuries. Nose injuries generally produce bleeding. The bleeding may be controlled by placing an ice pack (if available) over the nose, or pinching the nostrils together. The bleeding may also be controlled by placing torn gauze (rolled) between the upper teeth and the lip.

CAUTION

DO NOT attempt to remove objects inhaled into the nose. An untrained person who removes such an object could worsen the casualty's condition and cause permanent injury.

f. Jaw Injuries. Before applying a bandage to a casualty's jaw, remove all loose or free-floating foreign material from the casualty's mouth.

If the casualty is unconscious, check for obstructions in the airway and remove if possible. If there is profuse bleeding in the oral cavity, the cavity may require loose packing with soft bandaging material (for example: Kerlix™ gauze) prior to applying a bandage. Care should be taken to avoid occluding the airway. When applying the bandage, allow the jaw enough freedom to permit passage of air and drainage from the mouth.

(1) *Apply bandages attached to field first aid dressing to the jaw.* After dressing the wound, apply the bandages using the same technique illustrated in Figure 3-32A—C.

NOTE

The dressing and bandaging procedure outlined for the jaw serves a twofold purpose. In addition to stopping the bleeding and protecting the wound, it also immobilizes a fractured jaw.

(2) *Apply a cravat bandage to the jaw.*

(*a*) Place the bandage under the chin and pull its ends upward. Adjust the bandage to make one end longer than the other (Figure 3-32A).

(*b*) Take the longer end over the top of the head to meet the short end at the temple and cross the ends over (Figure 3-32B).

(*c*) Take the ends in opposite directions to the other side of the head and tie them over the part of the bandage that was applied first (Figure 3-32C).

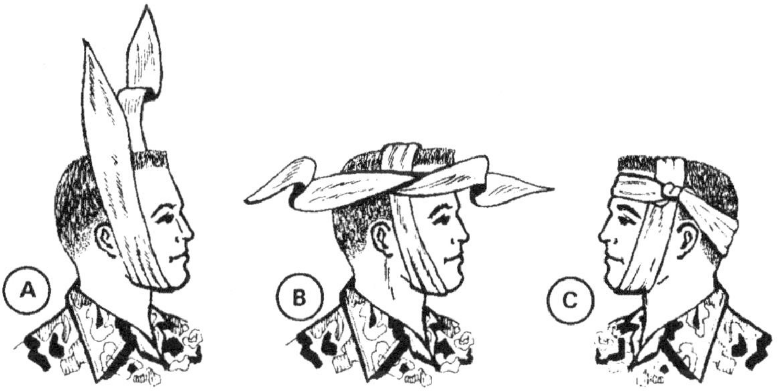

Figure 3-32. Applying a cravat bandage to jaw (Illustrated A—C).

NOTE

The cravat bandage technique is used to immobilize a fractured jaw or to maintain a sterile dressing that does not have tail bandages attached.

3-11. Shoulder Bandage

a. To apply bandages attached to the field first aid dressing—

(1) Take one bandage across the chest and the other across the back and under the arm opposite the injured shoulder.

(2) Tie the ends with a square knot (Figure 3-33).

Figure 3-33. Shoulder bandage.

b. To apply a cravat bandage to the shoulder or armpit—

(1) Make an extended cravat bandage by using two triangular bandages (Figure 3-34A); place the end of the first triangular bandage along the base of the second one (Figure 3-34B).

(2) Fold the two bandages into a single extended bandage (Figure 3-34C).

(3) Fold the extended bandage into a single cravat bandage (Figure 3-34D). After folding, secure the thicker part (overlap) with two or more safety pins (Figure 3-34E).

(4) Place the middle of the cravat bandage under the armpit so that the front end is longer than the back end and safety pins are on the outside (Figure 3-34F).

(5) Cross the ends on top of the shoulder (Figure 3-34G).

(6) Take one of the bandage ends across the back and under the arm on the opposite side and the other end across the chest. Tie the ends (Figure 3-34H).

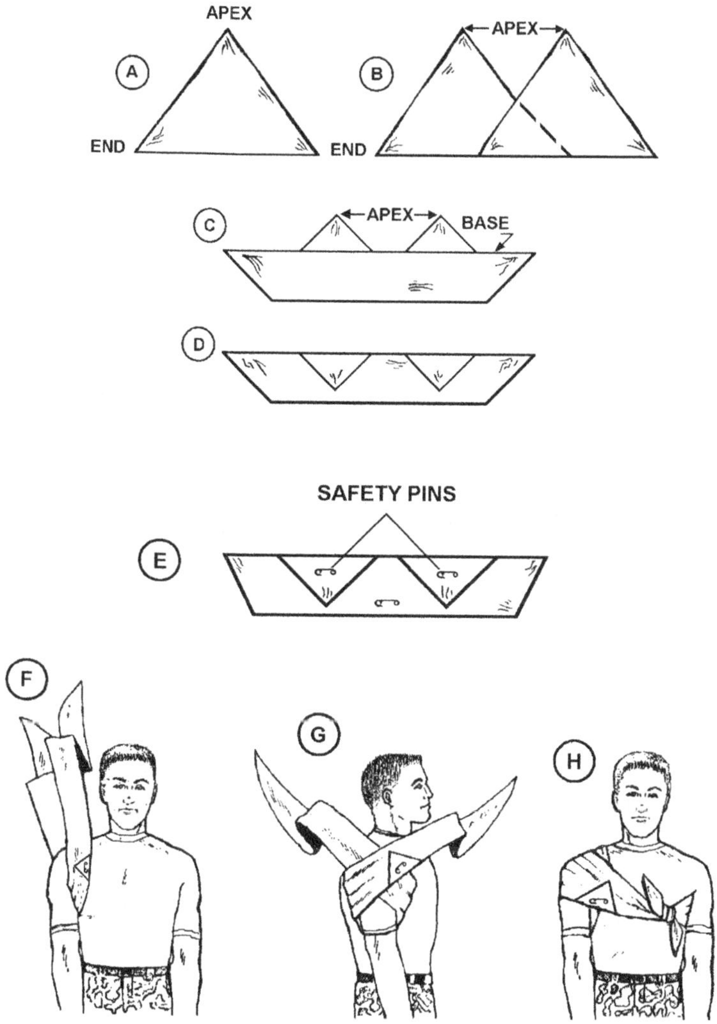

Figure 3-34. Extended cravat bandage applied to shoulder or armpit (Illustrated A—H).

Be sure to place sufficient wadding in the armpit. **DO NOT** tie the cravat bandage too tightly. Avoid compressing the major blood vessels in the armpit.

3-12. Elbow Bandage

To apply a cravat bandage to the elbow—

a. Bend the arm at the elbow and place the middle of the cravat at the point of the elbow bringing the ends upward (Figure 3-35A).

b. Bring the ends across, extending both downward (Figure 3-35B).

c. Take both ends around the arm and tie them with a square knot at the front of the elbow (Figure 3-35C).

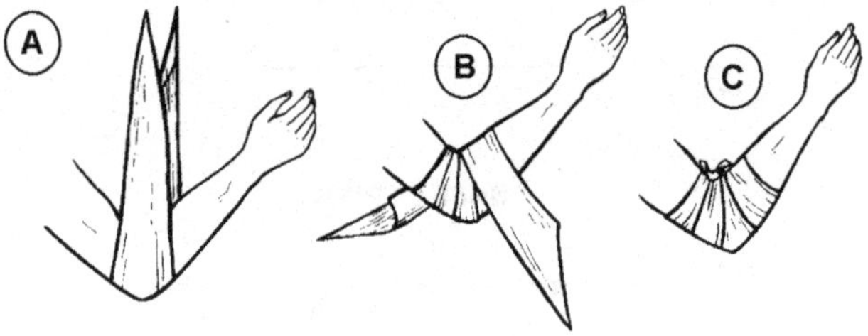

Figure 3-35. Elbow bandage (Illustrated A—C).

CAUTION
If an elbow fracture is suspected, **DO NOT** bend the elbow; bandage it in the position found.

3-13. Hand Bandage

a. To apply a triangular bandage to the hand—

(1) Place the hand in the middle of the triangular bandage with the wrist at the base of the bandage (Figure 3-36A). Ensure that the fingers are separated with absorbent material to prevent chafing and irritation of the skin.

(2) Place the apex over the fingers and tuck any excess material into the pleats on each side of the hand (Figure 3-36B).

(3) Cross the ends on top of the hand, take them around the wrist, and tie them (Figures 3-36C—E) with a square knot.

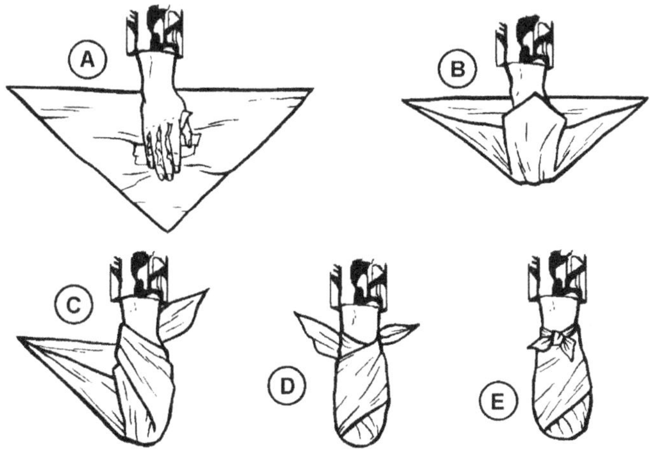

Figure 3-36. Triangular bandage applied to hand (Illustrated A—E).

b. To apply a cravat bandage to the palm of the hand—

(1) Lay the middle of the cravat over the palm of the hand with the ends hanging down on each side (Figure 3-37A).

(2) Take the end of the cravat at the little finger across the back of the hand, extending it upward over the base of the thumb; then bring it downward across the palm (Figure 3-37B).

(3) Take the thumb end across the back of the hand, over the palm, and through the hollow between the thumb and palm (Figure 3-37C).

(4) Take the ends to the back of the hand and cross them; then bring them up over the wrist and cross them again (Figure 3-37D).

(5) Bring both ends down and tie them with a square knot on top of the wrist (Figure 3-37E—F).

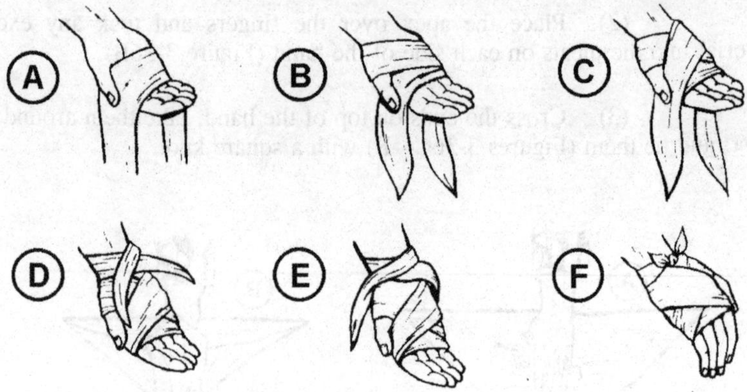

Figure 3-37. Cravat bandage applied to palm of hand (Illustrated A—F).

3-14. Leg (Upper and Lower) Bandage

To apply a cravat bandage to the leg—

a. Place the center of the cravat over the dressing (Figure 3-38A).

b. Take one end around and up the leg in a spiral motion and the other end around and down the leg in a spiral motion, overlapping part of each preceding turn (Figure 3-38B).

c. Bring both ends together and tie them (Figure 3-38C) with a square knot.

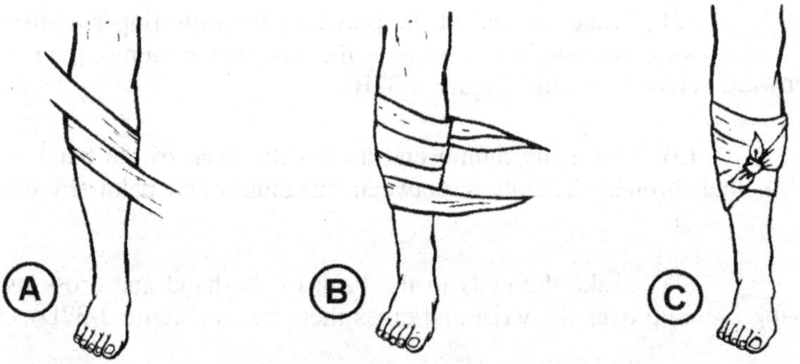

Figure 3-38. Cravat bandage applied to leg (Illustrated A—C).

3-15. Knee Bandage

To apply a cravat bandage to the knee as illustrated in Figure 3-39, use the same technique applied in bandaging the elbow.

> **CAUTION**
>
> If a fracture of the kneecap is suspected, **DO NOT** bend the knee; bandage it in the position found.

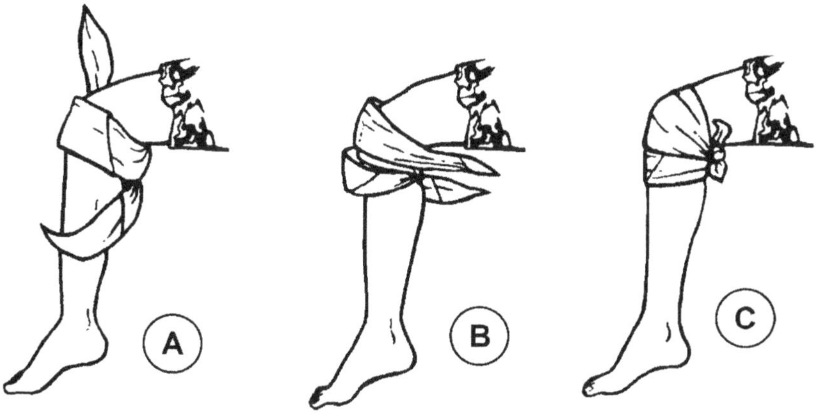

Figure 3-39. Cravat bandage applied to knee (Illustrated A—C).

3-16. Foot Bandage

To apply a triangular bandage to the foot—

a. Place the foot in the middle of the triangular bandage with the heel well forward of the base (Figure 3-40A). Ensure that the toes are separated by absorbent material to prevent chafing and irritation of the skin.

b. Place the apex over the top of the foot and tuck any excess material into the pleats on each side of the foot (Figure 3-40B).

c. Cross the ends on top of the foot, take them around the ankle, and tie them at the front of the ankle (Figure 3-40C—E).

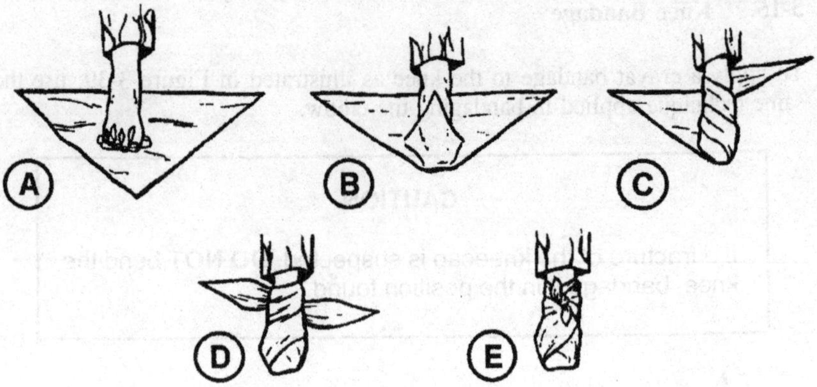

Figure 3-40. Triangular bandage applied to foot (Illustrated A—E).

CHAPTER 4

FIRST AID FOR FRACTURES

4-1. General

A fracture is any break in the continuity of a bone. Fractures can cause total disability or in some cases death by severing vital organs and/or arteries. On the other hand, they can most often be treated so there is a complete recovery. The potential for recovery depends greatly upon the first aid the individual receives before he is moved. First aid includes immobilizing the fractured part in addition to applying lifesaving measures when necessary. The basic splinting principle is to immobilize the joints above and below the fracture.

4-2. Kinds of Fractures

Figure 4-1 depicts types of fractures.

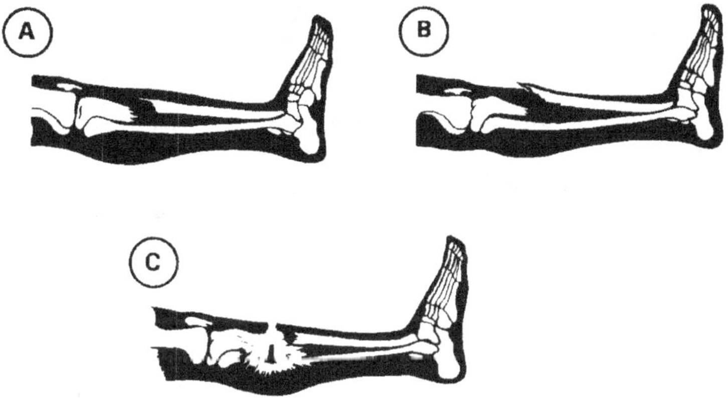

Figure 4-1. Types of fractures (*Illustrated A—C*).

a. Closed Fracture (Figure 4-1A). A closed fracture is a broken bone that does not break the overlying skin. The tissue beneath the skin may be damaged. A *dislocation* is when a joint, such as a knee, ankle, or shoulder, is not in the proper position. A *sprain* is when the connecting tissues of the joints have been torn. Dislocations and sprains (swelling, possible deformity, and discoloration) should be treated as closed fractures.

b. Open Fracture (Figure 4-1B and 4-1C). An open fracture is a broken bone that breaks (pierces) the overlying skin. The broken bone may come through the skin or a missile such as a bullet or shell fragment may go through the flesh and break the bone.

NOTE

An open fracture is contaminated and subject to infection.

4-3. Signs and Symptoms of Fractures

Indications of a fracture are deformity, tenderness, swelling, pain, inability to move the injured part, protruding bone, bleeding, or discolored skin at the injury site. A sharp pain when the service member attempts to move the part is also a sign of a fracture.

WARNING

DO NOT encourage the casualty to move the injured part in order to identify a fracture since such movement could cause further damage to surrounding tissues and promote shock. If you are not sure whether a bone is fractured, care for the injury as a fracture. At the site of the fracture, the bone ends are sharp and could cause vessel (artery and/or vein) damage.

4-4. Purposes of Immobilizing Fractures

A fracture is immobilized to prevent the sharp edges of the bone from moving and cutting tissue, muscle, blood vessels, and nerves. This reduces pain and helps prevent or control shock. In a closed fracture, immobilization keeps bone fragments from causing an open wound, which can become contaminated and subject to infection.

4-5. Splints, Padding, Bandages, Slings, and Swathes

a. *Splints*. Splints may be improvised from such items as boards, poles, sticks, tree limbs, or cardboard. If nothing is available for a splint, the chest wall can be used to immobilize a fractured arm and the uninjured leg can be used to immobilize (to some extent) the fractured leg.

b. *Padding*. Padding may be improvised from such items as a jacket, blanket, poncho, shelter half, or leafy vegetation.

c. Bandages. Bandages may be improvised from belts, rifle slings, kerchiefs, or strips torn from clothing or blankets. Narrow materials such as wire or cord should not be used to secure a splint in place. The application of wire and/or narrow material to an extremity could cause tissue damage and a tourniquet effect.

d. Slings. A sling is a bandage suspended from the neck to support an upper extremity. If a bandage is not available, a sling can be improvised by using the tail of a coat or shirt or pieces of cloth torn from such items as clothing and blankets. The triangular bandage is ideal for this purpose. Remember that the casualty's hand should be higher than his elbow, and the fingers should be showing at all times. The sling should be applied so that the supporting pressure is on the uninjured side.

e. Swathes. Swathes are any bands (pieces of cloth or load bearing equipment [LBE]) that are used to further immobilize a splinted fracture. Triangular and cravat bandages are often used and are called *swathe bandages*. The purpose of the swathe is to immobilize; therefore, the swathe bandage is placed above and/or below the fracture—not over it.

4-6. Procedures for Splinting Suspected Fractures

Before beginning first aid procedures for a fracture, gather whatever splinting materials are available. Ensure that splints are long enough to immobilize the joint above and below the suspected fracture. If possible, use at least four ties (two above and two below the fracture) to secure the splints. The ties should be square knots and should be tied away from the body on the splint. Distal pulses of the affected extremity should be checked before and after the application of the splint.

a. Evaluate the Casualty. Be prepared to perform any necessary lifesaving measures. Monitor the casualty for development of conditions that may require you to perform necessary lifesaving measures.

WARNING

Unless there is immediate life-threatening danger, such as a fire or an explosion, DO NOT move the casualty with a suspected back or neck injury. Improper movement may cause permanent paralysis or death.

WARNING

In a chemical environment, DO NOT remove any protective clothing. Apply the dressings and splints over the garments.

b. Locate the Site of the Suspected Fracture.

(1) Ask the casualty for the location of the injury.

- Does he have any pain?
- Where is it tender?
- Can he move the extremity?

NOTE

With the presence of an obvious deformity, do not make the casualty move extremity.

(2) Look for an unnatural position of the extremity.

(3) Look for a bone sticking out (protruding).

c. Prepare the Casualty for Splinting the Suspected Fracture.

(1) Reassure the casualty. Tell him that you will be providing first aid for him and that medical help is on the way.

(2) Loosen any tight or binding clothing.

(3) Remove all jewelry from the injured part and place it in the casualty's pocket. Tell the casualty you are doing this because if the jewelry is not removed and swelling occurs later, he may not be able to get it off and further bodily injury could result.

(4) Boots should not be removed from the casualty unless they are needed to stabilize a neck injury or there is actual bleeding from the foot.

d. Gather Splinting Materials. If standard splinting materials (splints, padding, and cravats) are not available, gather improvised materials. If splinting material is not available and the suspected fracture **CANNOT** be splinted, then swathes, or a combination of swathes and slings can be used to immobilize the extremity.

e. Pad the Splints. Pad the splints where they touch any bony part of the body, such as the elbow, wrist, knee, ankle, crotch, or armpit areas. Padding prevents excessive pressure on the area, which could lead to circulation problems.

f. Check the Circulation Below the Site of the Injury.

(1) Note any pale, white, or bluish-gray color of the skin, which may indicate impaired circulation. Circulation can also be checked by depressing the toe or fingernail beds and observing how quickly the color returns. A slower return of color to the injured side when compared with the uninjured side indicates a problem with circulation. The fingernail bed is the method to use to check the circulation in a dark-skinned casualty.

(2) Check the temperature of the injured extremity. Use your hand to compare the temperature of the injured side with the uninjured side. The body area below the injury may be colder to the touch indicating poor circulation.

(3) Question the casualty about the presence of numbness, tightness, cold, or tingling sensations.

WARNING

Casualties with fractures of the extremities may show impaired circulation, such as numbness, tingling, cold or pale to bluish skin tone. These casualties should be evacuated by medical personnel and treated as soon as possible. Prompt medical treatment may prevent possible loss of the limb.

WARNING

If it is an open fracture and the bone is protruding from the skin, *DO NOT ATTEMPT TO PUSH THE BONE BACK UNDER THE SKIN.* Apply a field dressing over the wound to protect the area.

g. Apply the Splint in Place.

(1) Splint the fracture in the position found. **DO NOT** attempt to reposition or straighten the injury. If it is an open fracture, stop the bleeding and protect the wound. Cover all wounds with field dressings before applying a splint. Remember to use the casualty's field dressing, not your own.

(2) Place one splint on each side of the fracture. Make sure that the splints reach, if possible, beyond the joints above and below the fracture.

(3) Tie the splints. Secure each splint in place above and below the fracture site with improvised (or actual) cravats. Improvised cravats, such as strips of cloth, belts, or whatever else you have, may be used. With minimal motion to the injured areas, place and tie the splints with the bandages. Push cravats through and under the natural body curvatures, and then gently position improvised cravats and tie in place. Use square knots. Tie all knots on the splint away from the casualty (Figure 4-2). **DO NOT** tie cravats directly over the suspected fracture site.

Figure 4-2. Square knots tied away from casualty.

h. Check the Splint for Tightness.

(1) **CHECK** to be sure that bandages are tight enough to securely hold splinting materials in place, but not so tight that circulation is impaired.

(2) **RECHECK** the circulation after application of the splint. Check the skin color and temperature. This is to ensure that the bandages holding the splint in place have not been tied too tightly. A fingertip check can be made by inserting the tip of the finger between the bandaged knot and the skin.

(3) **MAKE** any necessary adjustment without allowing the splint to become ineffective.

i. Apply a Sling. An improvised sling may be made from any available nonstretching piece of cloth, such as a battle dress uniform (BDU) shirt or trousers, poncho, or shelter half. Slings may also be improvised using the tail of a coat, belt, or a piece of cloth. Figure 4-3 depicts a shirttail used for support. A trousers belt or LBE may also be used for support (Figure 4-4). A sling should place the supporting pressure on the casualty's uninjured side. The supported arm should have the hand positioned slightly higher than the elbow showing the fingers.

Figure 4-3. Shirttail used for support.

Figure 4-4. Belt used for support.

(1) Insert the splinted arm in the center of the sling (Figure 4-5).

Figure 4-5. Arm inserted in center of improvised sling.

(2) Bring the ends of the sling up and tie them at the side (or hollow) of the neck on the uninjured side (Figure 4-6).

Figure 4-6. Ends of improvised sling tied to side of neck.

(3) Twist and tuck the corner of the sling at the elbow (Figure 4-7).

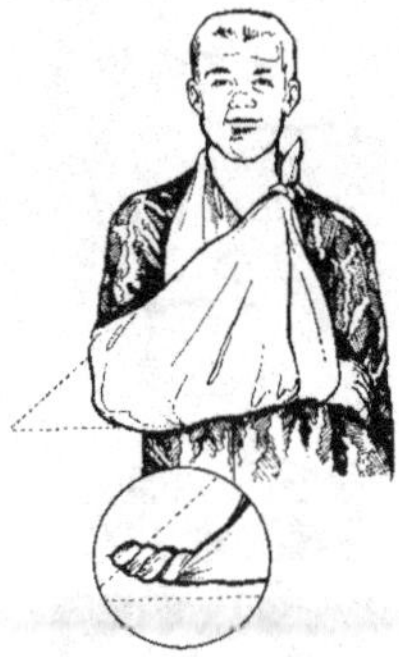

Figure 4-7. Corner of sling twisted and tucked at elbow.

j. Apply a Swathe. You may use any large piece of cloth, service member's belt, or pistol belt, to improvise a swathe.

WARNING

The swathe should not be placed directly on top of the injury, but positioned either above or below the fracture site.

(1) Apply swathes to the injured arm by wrapping the swathe over the injured arm, around the casualty's back, and under the arm on the uninjured side. Tie the ends on the uninjured side (Figure 4-8).

Figure 4-8. Arm immobilized with strip of clothing.

(2) A swathe is applied to an injured leg by wrapping the swathe around both legs and securing it on the uninjured side.

k. Seek Medical Assistance. Notify medical personnel, watch closely for development of life-threatening conditions and/or impaired circulation to the injured extremity. (Refer to Chapter 1 for additional information on life-threatening conditions.)

4-7. Upper Extremity Fractures

Figures 4-9 through 4-17 show how to apply slings, splints, and cravats (swathes) to immobilize and support fractures of the upper extremities. *Although the padding is not visible in some of the illustrations, it is always preferable to apply padding along the injured part for the length of the splint and especially where it touches any bony parts of the body.*

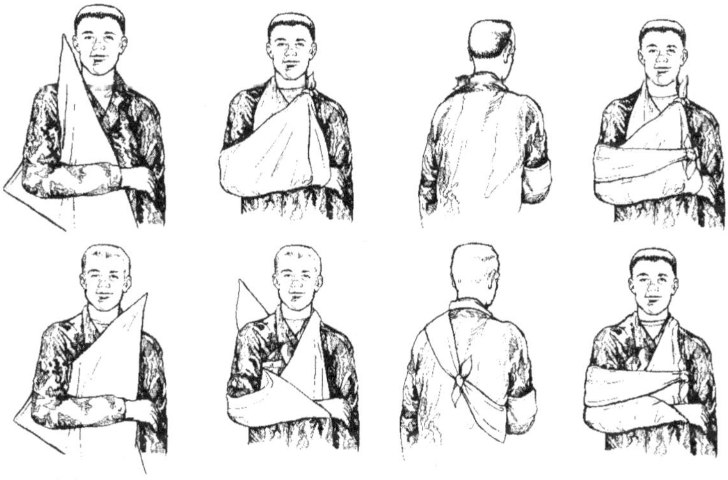

Figure 4-9. Application of triangular bandage to form sling (two methods).

Figure 4-10. Completing sling sequence by twisting and tucking the corner of the sling at the elbow.

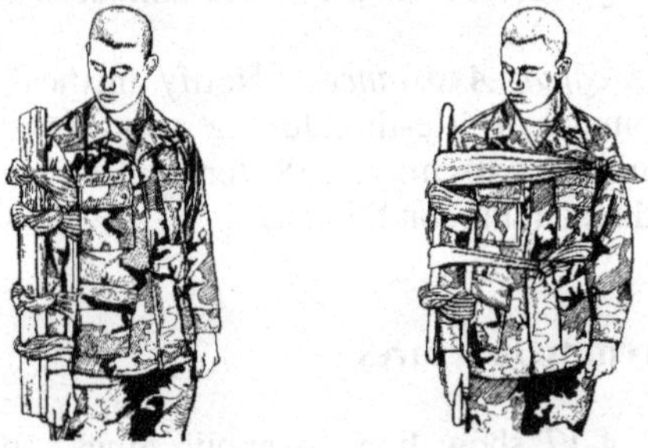

Figure 4-11. Board splints applied to fractured elbow when elbow is not bent (two methods).

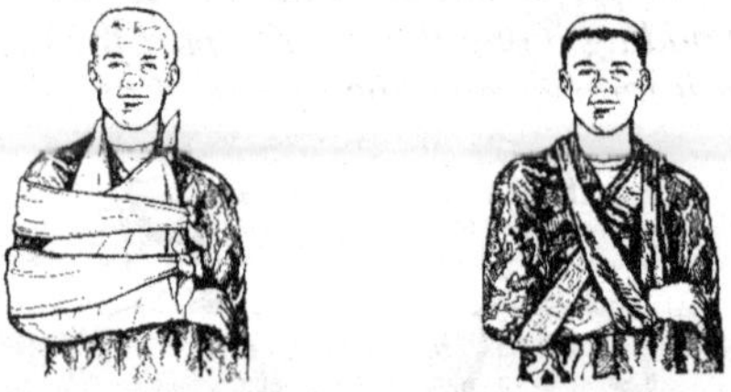

Figure 4-12. Chest wall used as splint for upper arm fracture when no splint is available.

Figure 4-13. Chest wall, sling, and cravat used to immobilize fractured elbow when elbow is bent.

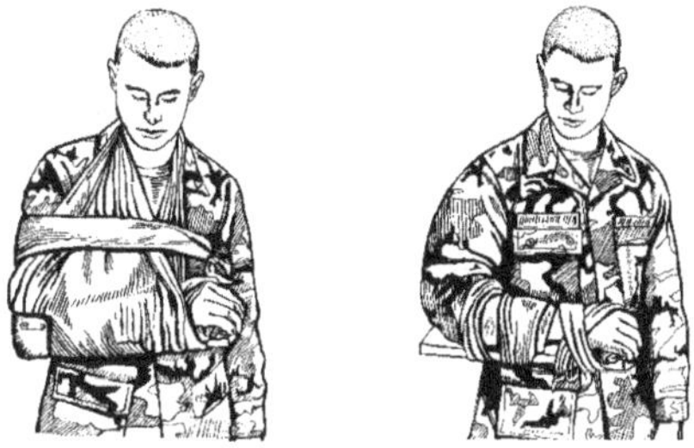

Figure 4-14. Board splint applied to fractured forearm.

Figure 4-15. Fractured forearm or wrist splinted with sticks and supported with tail of shirt and strips of material.

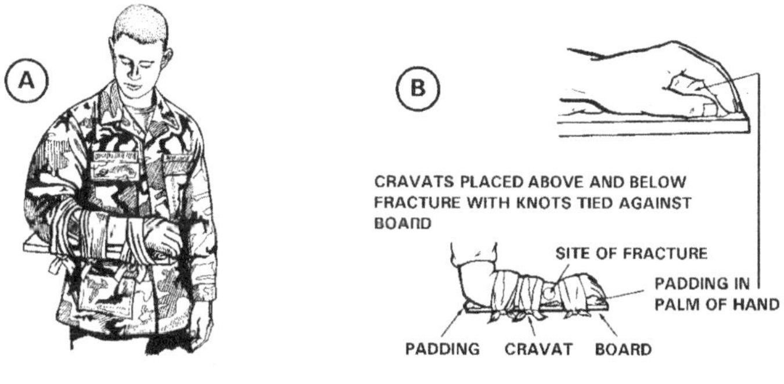

Figure 4-16. Board splint applied to fractured wrist and hand (Illustrated A—B).

Figure 4-17. SAM® splint applied to fractured wrist or forearm.

4-8. Lower Extremity Fractures

Figures 4-18 through 4-24 show how to apply splints to immobilize fractures of the lower extremities. *Although padding is not visible in some of the illustrations, it is always preferable to apply padding along the injured part for the length of the splint and especially where it touches any bony parts of the body.*

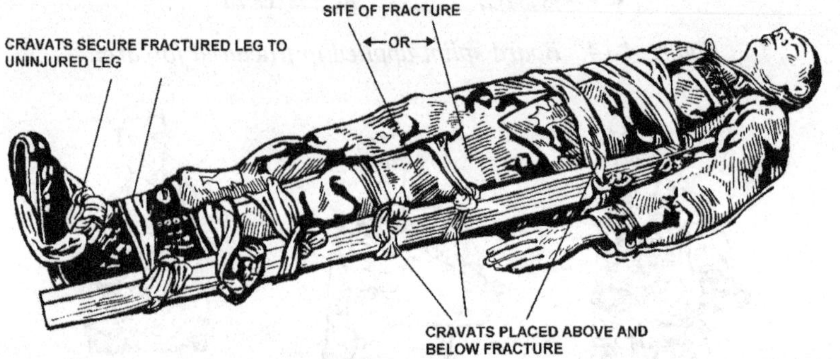

Figure 4-18. Board splints applied to fractured hip or thigh.

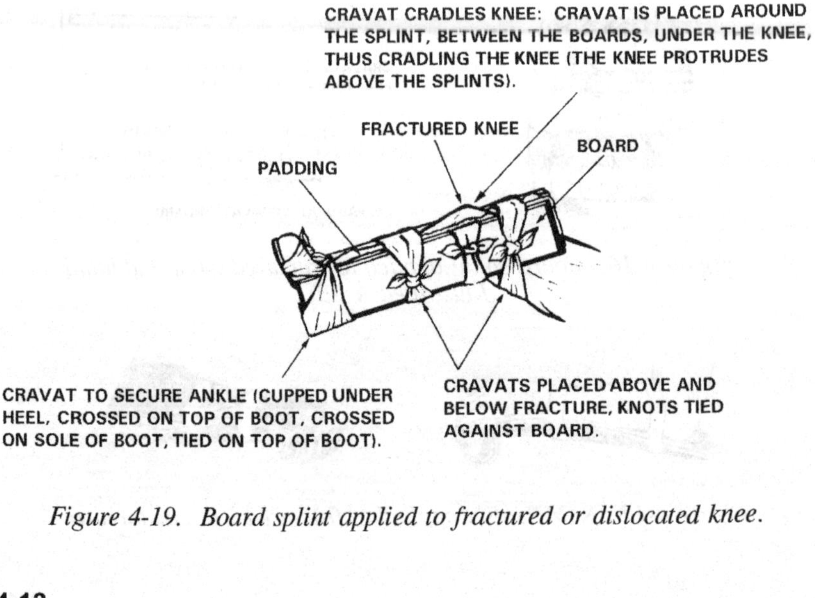

Figure 4-19. Board splint applied to fractured or dislocated knee.

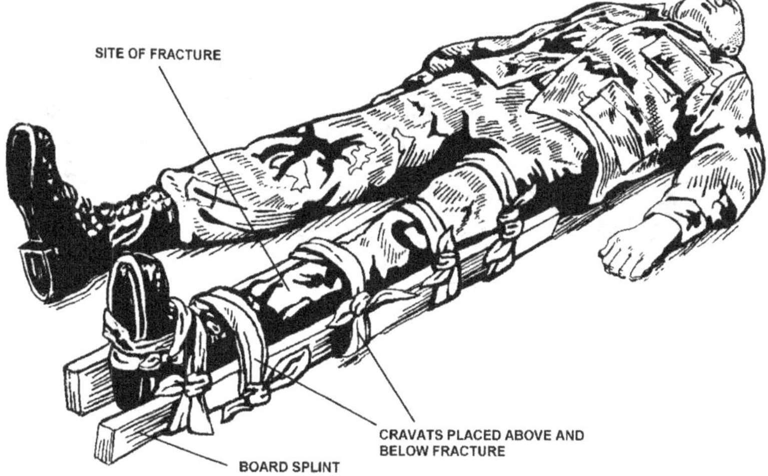

Figure 4-20. Board splints applied to fractured lower leg or ankle.

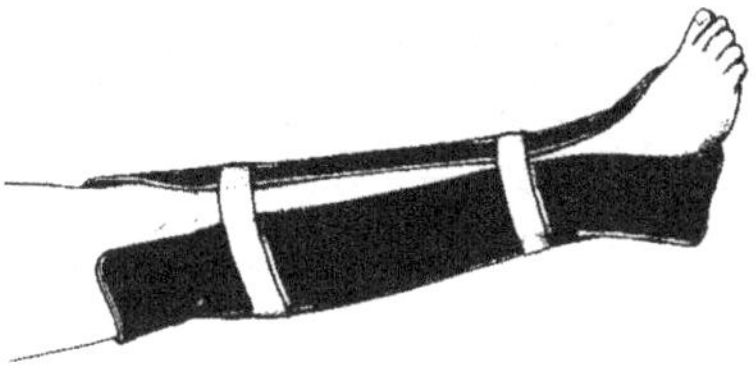

Figure 4-21. SAM® splint applied to fractured lower leg or ankle.

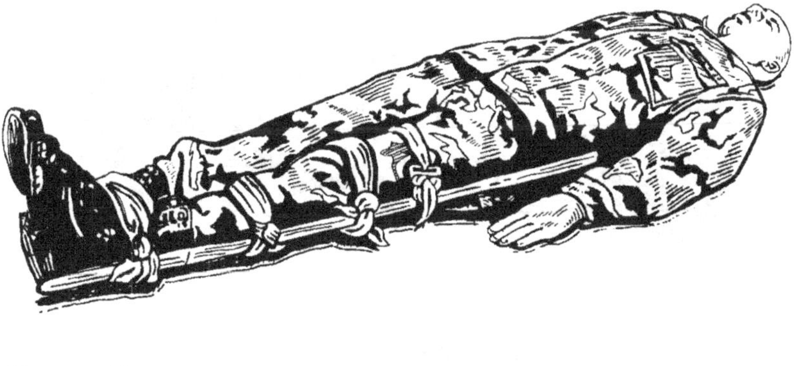

Figure 4-22. Improvised splints applied to fractured lower leg or ankle.

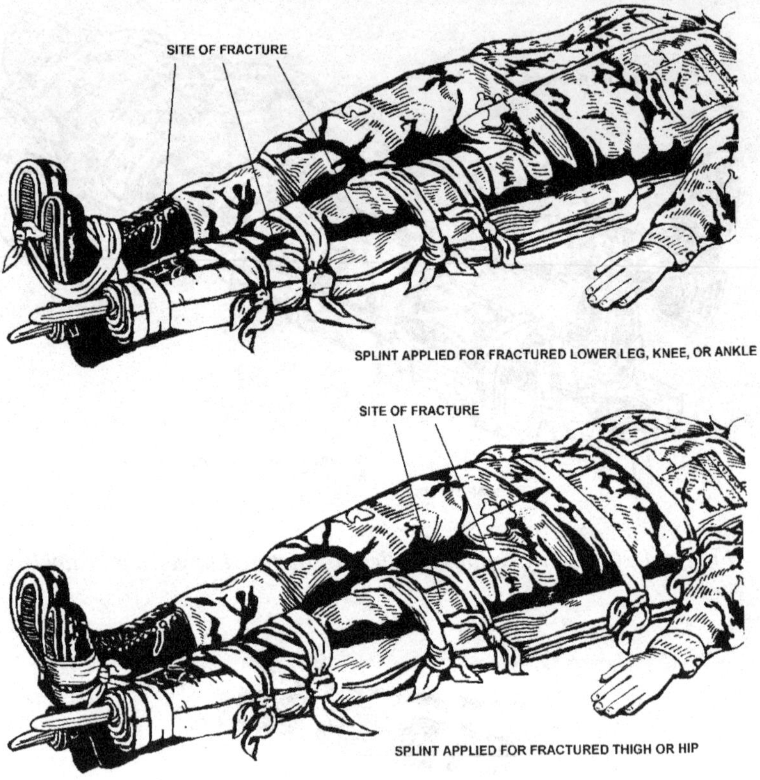

Figure 4-23. Poles rolled in a blanket and used as splints applied to fractured lower extremity.

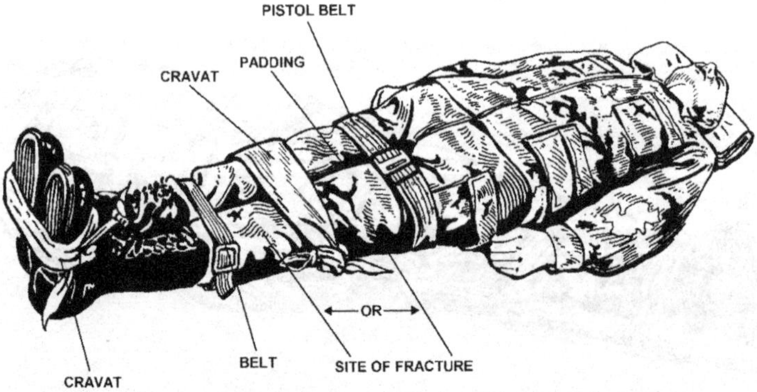

Figure 4-24. Uninjured leg used as splint for fractured leg (anatomical splint).

4-9. Jaw, Collarbone, and Shoulder Fractures

a. Apply a cravat to immobilize a fractured jaw as illustrated in Figure 4-25. Direct all bandaging support to the top of the casualty's head, not to the back of his neck. If incorrectly placed, the bandage will pull the casualty's jaw back and interfere with his breathing.

Figure 4-25. Fractured jaw immobilized.

> **WARNING**
>
> **Casualties with lower jaw (mandible) fractures cannot be laid flat on their backs because facial muscles will relax and may cause an airway obstruction.**

b. Apply two belts, a sling, and a cravat to immobilize a fractured collarbone, as illustrated in Figure 4-26.

Figure 4-26. Application of belts, sling, and cravat to immobilize a fractured collarbone.

c. Apply a sling and a cravat to immobilize a fractured or dislocated shoulder, using the technique illustrated in Figure 4-27.

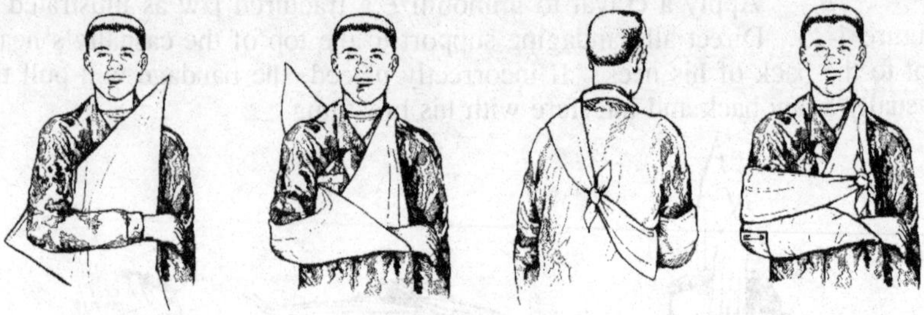

Figure 4-27. Application of sling and cravat to immobilize a fractured or dislocated shoulder.

4-10. Spinal Column Fractures

It is often impossible to be sure a casualty has a fractured spinal column. Be suspicious of any back injury, especially if the casualty has fallen or if his back has been sharply struck or bent. If a casualty has received such an injury and does not have feeling in his legs or cannot move them, you can be reasonably sure that he has a severe back injury, which should be managed as a fracture. Remember, that the possibility of a neck fracture or injury to the back should always be suspected, and it is often impossible to be sure if a casualty has a fractured spinal column. If the spine is fractured, bending it can cause the sharp bone fragments to bruise or cut the spinal cord and result in permanent paralysis or death (Figure 4-28A). The spinal column must maintain normal spinal position at the lower back (lumbar region) to help remove pressure from the spinal cord.

a. If the casualty is not to be transported until medical personnel arrive—

- Caution him not to move. Ask him if he is in pain or if he is unable to move any part of his body.

- Leave him in the position in which he is found. **DO NOT** move any part of his body, unless he is in imminent danger.

- If the casualty is lying face up, slip a blanket or other supporting material under the arch of his lower back to help support the spine in a normal position (Figure 4-28B). Take care not to place so much bulky padding as to cause potential damage by causing undo pressure on the spine. If he is lying face down, **DO NOT** put anything under any part of his body.

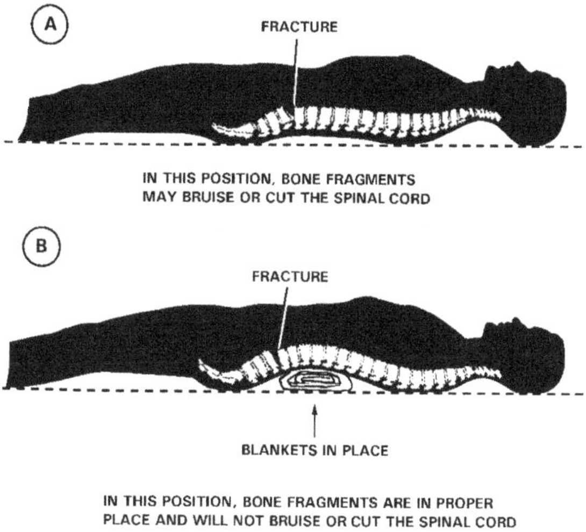

Figure 4-28. Spinal column must maintain a normal spine position.

b. If the casualty must be transported to a safe location before medical personnel arrive and if the casualty is in a—

- Face-up position, transport him by litter or use a firm substitute, such as a wide board or a door longer than his height. Loosely tie the casualty's wrists together over his waistline, using a cravat or a strip of cloth. Tie his feet together to prevent the accidental dropping or shifting of his legs. Lay a folded blanket across the litter where the arch of his back is to be placed. Using a four-man team (Figure 4-29), place the casualty on the litter without bending his spinal column or his neck.

Figure 4-29. Placing face-up casualty with fractured back onto litter.

- The number two man positions himself at the casualty's head. His responsibility is to provide manual in-line (neutral) stabilization of the head and neck. The number three, and four men position themselves on one side of the casualty; all kneel on one knee along the side of the casualty. The number one man positions himself to the opposite side of the casualty (or can be on the same side of number three and four). The numbers two, three, and four men gently place their hands under the casualty. The number one man on the opposite side places his hands under the injured part to assist.

- When all four men are in position to lift, the number two man commands, "**PREPARE TO LIFT**" and then, "**LIFT**." All men, in unison, gently lift the casualty about 8 inches. Once the casualty is lifted, the number one man recovers and slides the litter under the casualty, ensuring that the blanket is in proper position. The number one man then returns to his original lift position (Figure 4-29).

- When the number two man commands, "**LOWER CASUALTY**," all men, in unison, gently lower the casualty onto the litter.

- Facedown position, he must be transported in this same position. The four-man team lifts him onto a regular or improvised litter, keeping the spinal column in a normal spinal position. If a regular litter is used, first place a folded blanket on the litter at the point where the chest will be placed.

4-11. Neck Fractures

A fractured neck is extremely dangerous. Bone fragments may bruise or cut the spinal cord just as they might in a fractured back.

a. If the casualty is not to be transported until medical personnel arrive—

- Caution him not to move. Moving may cause permanent injury or death.

- Leave the casualty in the position in which he is found. If his neck and head (cervical spine) are in an abnormal position, immediately immobilize the neck and head.

- Keep his head still, if the casualty is lying face up, raise his shoulders slightly, and slip a roll of cloth that has the bulk of a bath towel under his neck (Figure 4-31). The roll should be thick enough to arch his neck only slightly, leaving the back of his head on the ground. **DO NOT** bend his neck or head forward. **DO NOT** raise or twist his head. Immobilize the casualty's head (Figure 4-32). Do this by padding heavy objects (such as rocks or his boots filled with dirt, sand, gravel, or rock) and placing them on each side of his head. If it is necessary to use boots, after filling them, tie the top tightly or stuff with pieces of cloth to secure the contents.)

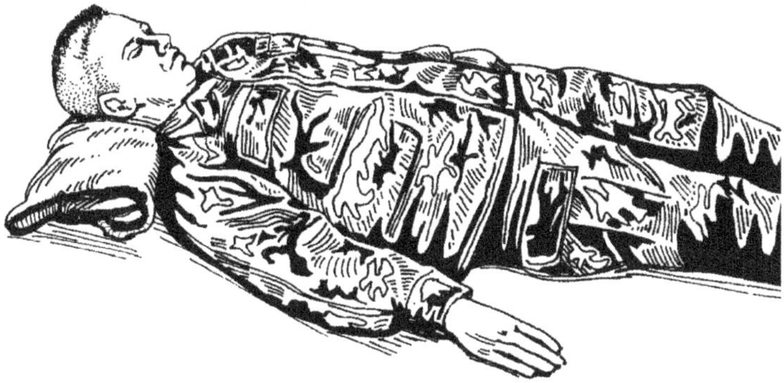

Figure 4-30. Casualty with roll of cloth (bulk) under neck.

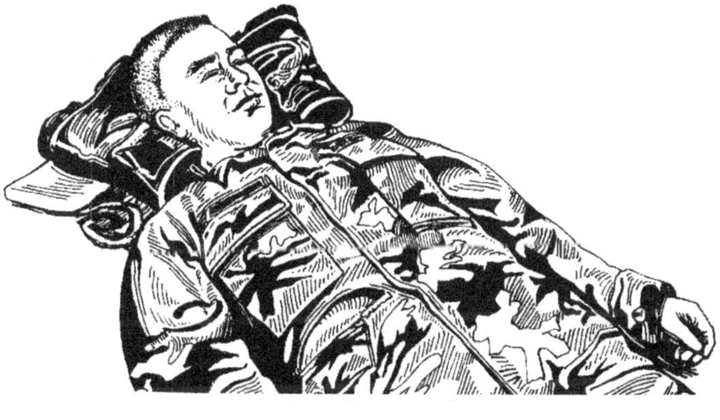

Figure 4-31. Immobilization of fractured neck.

- **DO NOT** move him if the casualty is lying face down. Immobilize the head and neck by padding heavy objects and placing them on each side of his head. DO NOT put a roll of cloth under the neck. **DO NOT** bend the neck or head, nor roll the casualty onto his back.

b. If the casualty must be prepared for transportation before medical personnel arrive—

- If the casualty has a fractured neck, at least two persons are needed because the casualty's head and trunk must be moved in unison. The two persons must work in close coordination (Figure 4-32) to avoid bending of the neck.

- A wide board is placed lengthwise beside the casualty. It should extend at least 4 inches beyond the casualty's head and feet (Figure 4-32A).

- If the casualty is lying face up, the number one man steadies the casualty's head and neck between his hands. At the same time, the number two man positions one foot and one knee against the board to prevent it from slipping. He then grasps the casualty underneath his shoulder and hip and gently slides him onto the board (Figure 4-32B).

- If the casualty is lying face down, the number one man steadies the casualty's head and neck between his hands, while the number two man gently rolls the casualty over onto the board (Figure 4-32C).

- The number one man continues to steady the casualty's head and neck. The number two man simultaneously raises the casualty's shoulders slightly, places padding under his neck, and immobilizes the casualty's head (Figures 4-32D—E).

- Any improvised supports are secured in position with a cravat or strip of cloth extended across the casualty's forehead and under the board (Figure 4-32D).

- The board is lifted onto a litter or blanket in order to transport the casualty (Figure 4-32E).

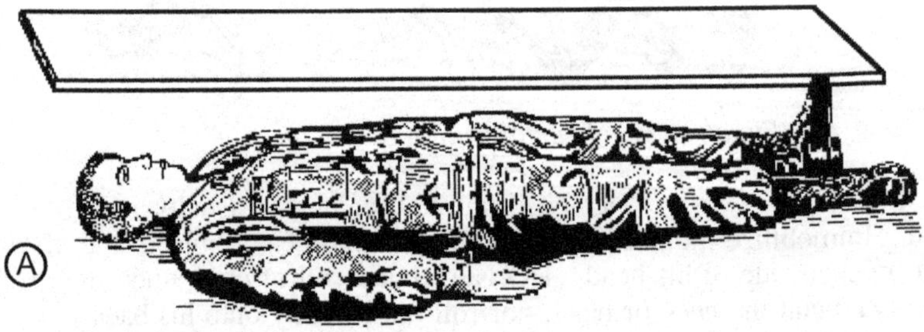

Figure 4-32. Preparing casualty with fractured neck for transportation (Illustrated A—E).

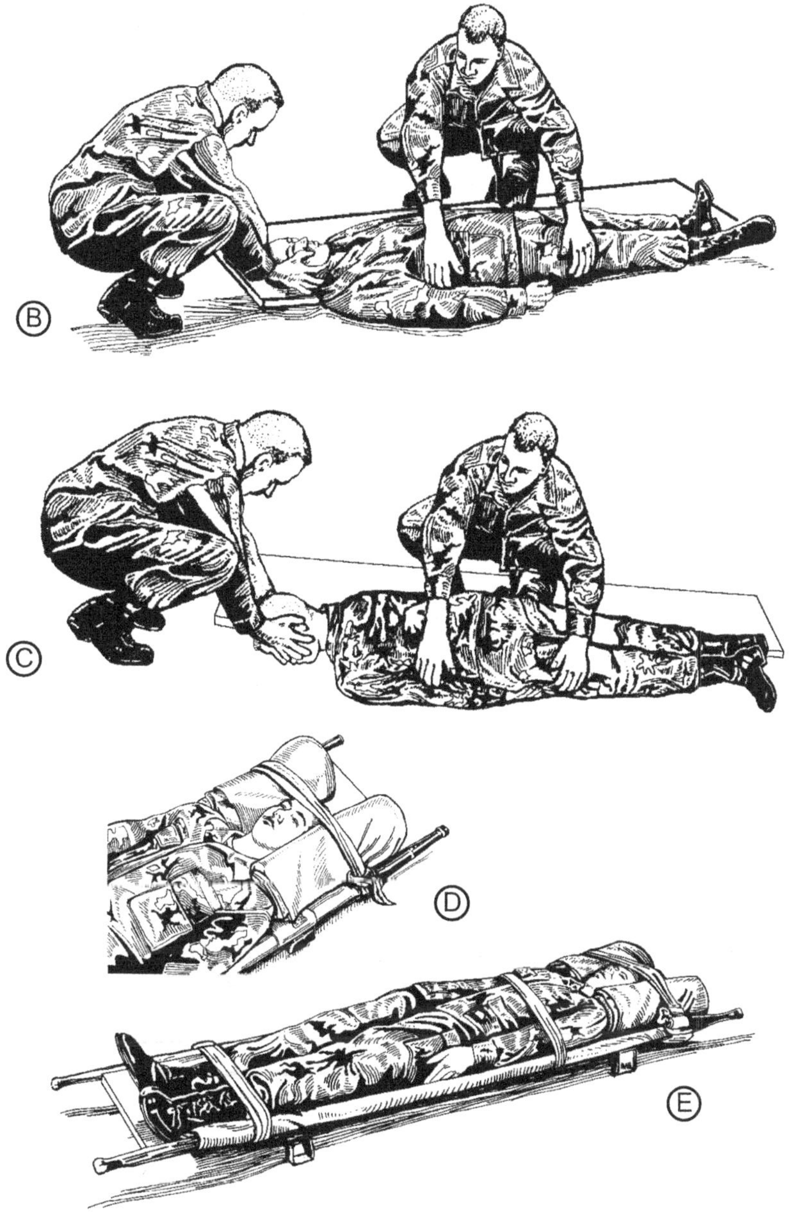

Figure 4-32. Preparing casualty with fractured neck for transportation (Illustrated A—E) (Continued).

CHAPTER 5

FIRST AID FOR CLIMATIC INJURIES

5-1. General

a. It is desirable, but not always possible, for an individual's body to become adjusted (acclimated) to an environment.

(1) The service members physical condition determines the amount of time their bodies need to adjust to the environment. Even those individuals in good physical condition need time before working or training in extremes of hot or cold weather. Climate-related injuries are usually preventable; prevention is both an individual and leadership responsibility.

(2) Several factors contribute to health and well-being in any environment—

- Diet.
- Sleep and rest.
- Exercise.
- Suitable clothing.

(3) Diet should be suited to an individual's needs in a particular climate. A special diet started for any purpose (such as weight reduction) should be done with appropriate medical supervision.

WARNING

Service members should use extreme caution when starting fad diets or taking over-the-counter herbal supplements. Medical records revealed that deaths and severe injuries occurred in individuals using dietary/ herbal supplements without medical monitoring.

NOTE

Weight loss and the use of weight loss supplements should be supervised by a trained health care provider.

(4) Specialized clothing and equipment (such as cold weather gear) for a specific environment should be obtained and used properly.

b. For information on the prevention of heat and cold injuries, refer to FM 21-10/Marine Corps Reference Publication (MCRP) 4-11.1D.

5-2. Heat Injuries

a. Heat injuries are environmental injuries. They may result when a service member—

- Is exposed to extreme heat, such as from the sun or from high temperatures.
- Does not wear proper clothing.
- Is in MOPP gear.
- Is inside closed spaces, such as inside an armored vehicle.
- Wears body armor.

b. Heat injury can be divided into three categories: heat cramps, heat exhaustion, and heatstroke.

c. Each service member must be able to recognize and give first aid for heat injuries.

WARNING

The heat casualty should be continually monitored for development of conditions which may require the performance of necessary basic lifesaving measures.

CAUTION

Do not use salt solutions in first aid procedures for heat injuries.

(1) Check the casualty for signs and symptoms of *cramping*.

- *Signs and symptoms.* Cramping is caused by an imbalance of chemicals (called electrolytes) in the body as a result of excessive sweating. This condition causes the casualty to exhibit:

- Cramping in the extremities (arms and legs).
- Abdominal (stomach) cramps.
- Excessive sweating.

NOTE

Thirst may or may not occur. Cramping can occur without the service member being thirsty.

- *First aid measures.*
  - Move the casualty to a cool, shady area or improvise shade if none is available.
  - Loosen his clothing (if not in a chemical environment).

NOTE

In a chemical environment, transport the heat casualty to a noncontaminated area as soon as the mission permits.

  - Have him slowly drink at least one canteen full of water. (The body absorbs cool water faster than warm or cold water; therefore, cool water is preferred if it is available.)
  - Seek medical assistance should cramps continue.

(2) Check the casualty for signs and symptoms of *heat exhaustion*.

- *Signs and symptoms.* Heat exhaustion is caused by loss of body fluids (dehydration) through sweating without adequate fluid replacement. It can occur in an otherwise fit individual who is involved in physical exertion in any hot environment especially if the service member is not acclimatized to that environment. These signs and symptoms are—
  - Excessive sweating with pale, moist, cool skin.
  - Headache.
  - Weakness.

- Dizziness.
- Loss of appetite.
- Cramping.
- Nausea (with or without vomiting).
- Urge to defecate.
- Chills (gooseflesh).
- Rapid breathing.
- Tingling of hands and/or feet.
- Confusion.

- *First aid measures.*

    - Move the casualty to a cool, shady area or improvise shade if none is available.
    - Loosen or remove his clothing and boots (unless in a chemical environment); pour water on him and fan him.
    - Have him slowly drink at least one canteen of water.
    - Elevate his legs.
    - If possible, the casualty should not participate in strenuous activity for the remainder of the day.
    - Monitor the casualty until the symptoms are gone, or medical assistance arrives.

(3) Check the casualty for signs and symptoms of *heatstroke*.

WARNING

Heatstroke is a medical emergency which may result in death if care is delayed.

- *Signs and symptoms.* A service member suffering from heatstroke has been exposed to high temperatures (such as direct sunlight) or been dressed in protective overgarments, which causes the body temperature to rise. Heatstroke occurs more rapidly in service members who are engaged in work or other physical activity in a high heat environment. Heatstroke is caused by a failure of the body's cooling mechanism which includes a decrease in the body's ability to produce sweat. The casualty's skin is red (flushed), hot, and dry. He may experience weakness, dizziness, confusion, headaches, seizures, nausea, stomach pains or cramps, and his respiration and pulse may be rapid and weak. Unconsciousness and collapse may occur suddenly.

- *First aid measures.* Cool casualty immediately by—

  - Moving him to a cool, shady area or improvising shade if none is available.

  - Loosening or removing his clothing (except in a chemical environment).

  - Spraying or pouring water on him; fanning him to permit the coolant effect of evaporation.

  - Massaging his extremities and skin, which increases the blood flow to those body areas, thus aiding the cooling process.

  - Elevating his legs.

  - Having him slowly drink at least one canteen full of water if he is conscious.

NOTE

Start cooling casualty immediately. Continue cooling while awaiting transportation and during transport to an MTF.

- *Medical assistance.* Seek medical assistance because the casualty should be transported to an MTF as soon as possible. Do not interrupt the cooling process or lifesaving measures to seek help; if someone else is present send them for help. The casualty should be continually monitored for development of conditions that may require the performance of necessary basic lifesaving measures.

d. *Table.* See Table 5-1 for further information.

Table 5-1. Heat Injuries

INJURIES	SIGNS AND SYMPTOMS	FIRST AID[1]
HEAT CRAMPS	THE CASUALTY EXPERIENCES MUSCLE CRAMPS OF THE ARMS, LEGS, AND/OR STOMACH. THE CASUALTY MAY ALSO HAVE EXCESSIVE SWEATING.	1. MOVE THE CASUALTY TO A COOL SHADY AREA OR IMPROVISE SHADE AND LOOSEN CLOTHING.[2] 2. HAVE HIM SLOWLY DRINK AT LEAST ONE CANTEEN FULL OF COOL WATER SLOWLY. 3. MONITOR THE CASUALTY AND GIVE HIM MORE WATER AS TOLERATED.
HEAT EXHAUSTION	THE CASUALTY EXPERIENCES HEAVY SWEATING WITH PALE, MOIST, COOL SKIN; HEADACHE, WEAKNESS, DIZZINESS, AND/OR LOSS OF APPETITE, HEAT CRAMPS, NAUSEA (WITH OR WITHOUT VOMITING), URGE TO DEFECATE, CHILLS (GOOSE-FLESH), RAPID BREATHING, CONFUSION, AND TINGLING OF THE HANDS AND/OR FEET.	1. MOVE THE CASUALTY TO A COOL, SHADY AREA OR IMPROVISE SHADE AND LOOSEN OR REMOVE HIS CLOTHING.[2] 2. POUR WATER ON HIM AND FAN HIM TO PERMIT THE COOLANT EFFECT OF EVAPORATION. 3. HAVE HIM SLOWLY DRINK AT LEAST ONE CANTEEN FULL OF COOL WATER. 4. ELEVATE THE CASUALTY'S LEGS. 5. SEEK MEDICAL ASSISTANCE IF SYMPTOMS CONTINUE; MONITOR UNTIL SYMPTOMS ARE GONE OR MEDICAL ASSISTANCE ARRIVES.
HEATSTROKE[3] (SUNSTROKE)	THE CASUALTY STOPS SWEATING (RED [FLUSHED] HOT, DRY SKIN). HE FIRST MAY EXPERIENCE HEADACHE, DIZZINESS, NAUSEA, FAST PULSE AND RESPIRATION, SEIZURES, AND MENTAL CONFUSION. HE MAY COLLAPSE	1. MOVE THE CASUALTY TO A COOL, SHADY AREA OR IMPROVISE SHADE AND LOOSEN OR REMOVE HIS CLOTHING, REMOVE THE OUTER GARMENTS AND PROTECTIVE

Table 5-1. Heat Injuries (Continued)

INJURIES	SIGNS AND SYMPTOMS	FIRST AID[1]
	AND SUDDENLY BECOME UNCONSCIOUS. **THIS IS A MEDICAL EMERGENCY.**	CLOTHING IF THE SITUATION PERMITS.[2] 2. START COOLING THE CASUALTY IMMEDIATELY. SPRAY OR POUR WATER ON HIM. FAN HIM. MASSAGE HIS EXTREMITIES AND SKIN. 3. ELEVATE HIS LEGS. 4. IF CONSCIOUS, HAVE HIM SLOWLY DRINK AT LEAST ONE CANTEEN FULL OF COOL WATER. 5. SEEK MEDICAL AID. CONTINUE COOLING WHILE AWAITING TRANSPORT AND CONTINUE FIRST AID WHILE EN ROUTE.

LEGEND:

1 THE FIRST AID PROCEDURE FOR HEAT RELATED INJURIES CAUSED BY WEARING INDIVIDUAL PROTECTIVE EQUIPMENT (IPE) IS TO MOVE THE CASUALTY TO A CLEAN AREA AND GIVE HIM WATER TO DRINK.
2 WHEN IN A CHEMICAL ENVIRONMENT, DO NOT LOOSEN OR REMOVE THE CASUALTY'S CLOTHING.
3 CAN BE FATAL IF NOT PROVIDED FIRST AID AND MEDICAL TREATMENT PROMPTLY.

5-3. Cold Injuries

Cold injuries are most likely to occur when conditions are moderately cold, but accompanied by wet or windy conditions. Cold injuries can usually be prevented. Well-disciplined and well-trained service members can be protected even in the most adverse circumstances. They and their leaders must know the hazards of exposure to the cold. They must know the importance of personal hygiene, exercise, care of the feet and hands, and the use of protective clothing.

a. Contributing Factors.

(1) Temperature, humidity, precipitation, and wind greatly increase likelihood of cold injuries, and the service members with wet clothing are at great risk of cold injuries. Riverine operations (river, swamp, and stream crossings) increase likelihood of cold injuries. Low temperatures and low relative humidity (dry cold) promote frostbite. Higher temperatures, together with moisture, promote immersion syndrome. Windchill accelerates the loss of body heat and may aggravate cold injuries.

(2) Relatively stationary activities such as being in an observation post or on guard duty increase the service member's vulnerability to cold injury. Also, a service member is more likely to receive a cold injury if he is—

- In contact with the ground (such as marching, performing guard duty, or engaging in other outside activities).
- Immobile for long periods (such as while riding in an unheated or open vehicle).
- Standing in water, such as in a foxhole.
- Out in the cold for days without being warmed.
- Deprived of an adequate diet and rest.
- Not able to take care of his personal hygiene.

(3) Physical fatigue contributes to apathy, which leads to inactivity, personal neglect, carelessness, and reduced heat production. In turn, these increase the risk of cold injury. Service members with prior cold injuries have a higher-than-normal risk of subsequent cold injury; not necessarily involving the body part previously injured.

(4) Depressed or unresponsive service members are also vulnerable because they are less active. These service members tend to be careless about precautionary measures, especially warming activities, when cold injury is a threat.

(5) Excessive use of alcohol or drugs leading to faulty judgment or unconsciousness in a cold environment increases the risk of becoming a cold injury casualty.

b. Signs and Symptoms. Once a service member becomes familiar with the factors that contribute to cold injury, he must learn to recognize cold injury signs and symptoms.

(1) Many service members suffer cold injury without realizing what is happening to them. They may be cold and generally uncomfortable. These service members often do not notice the injured part because it is already numb from the cold.

(2) Superficial cold injury usually can be detected by numbness or tingling sensations. These signs and symptoms often can be relieved simply by loosening boots or other clothing and by exercising to improve circulation. In more advanced cases involving deep cold injury, the service member often is not aware that there is a problem until the affected part feels like a stump or block of wood.

(3) Outward signs of cold injury include discoloration of the skin at the site of injury. In light-skinned persons, the skin first reddens and then becomes pale or waxy white. In dark-skinned persons, grayness in the skin is usually evident. An injured foot or hand feels cold to the touch. Swelling may be an indication of deep injury. Also note that blisters may occur after rewarming the affected parts. Service members should work in pairs (buddy teams) to check each other for signs of discoloration and other symptoms.

c. *First Aid Measures.* First aid for cold injuries depends on whether they are superficial or deep. Rewarming the affected part using body heat can adequately treat cases of superficial cold injury. (For example, this can be done by covering cheeks with hands, putting fingertips in armpits, or placing the casualty's feet under the clothing of a buddy [next to his belly].) The injured part should **NOT** be massaged, exposed to a fire or stove, rubbed with snow, slapped, chafed, or soaked in cold water. Walking on injured feet should be avoided. Deep cold injury (frostbite) is very serious and requires prompt first aid to avoid or to minimize the loss of parts or all of the fingers, toes, hands, or feet. The sequence for treating cold injuries depends on whether the condition is life-threatening. The first priority in managing cold injuries is to remove the casualty from the cold environment (such as building an improvised shelter). Other injuries the casualty may have are provided first aid simultaneously while waiting for transportation or evacuation. If the casualty is to be transported in a nonmedical vehicle, first aid measures should be continued en route to the MTF.

d. *Conditions Caused by Cold.* Conditions caused by cold include chilblain, immersion syndrome (immersion foot and trench foot), frostbite, snow blindness, dehydration, and hypothermia.

(1) *Chilblain.*

- *Signs and symptoms.* Chilblain is caused by repeated prolonged exposure of bare skin at temperatures from 60° Fahrenheit (F) to

32°F, or 20°F for acclimated, dry, unwashed skin. The area may be acutely swollen, red, tender, and hot with itchy skin. There may be no loss of skin tissue in untreated cases but continued exposure may lead to infected, ulcerated, or bleeding lesions.

• *First aid measures.* Within minutes, the area usually responds to locally applied body heat. Rewarm the affected part by applying firm steady pressure with your hands, or placing the affected part under your arms or against the stomach of a buddy. **DO NOT** rub or massage affected areas.

NOTE

Medical personnel should evaluate the injury, because signs and symptoms of tissue damage may be slow to appear.

(2) *Immersion syndrome (immersion foot and trench foot).* Immersion foot and trench foot are injuries that result from fairly long exposure of the feet to wet conditions at temperatures from approximately 32°F to 50°F. Inactive feet in damp or wet socks and boots, or tightly laced boots which impair circulation, are even more susceptible to injury. This injury can be very serious; it can lead to loss of toes or parts of the feet. If exposure of the feet has been prolonged and severe, the feet may swell so much that pressure closes the blood vessels and cuts off circulation. Should an immersion injury occur, dry the feet thoroughly and transport the casualty to an MTF by the fastest means possible.

• *Signs and symptoms.* At first, the parts of the affected foot are cold and painless, the pulse is weak, and numbness may be present. Second, the parts may feel hot, and burning and shooting pains may begin. In later stages, the skin is pale with a bluish cast and the pulse decreases. Other signs and symptoms that may follow are blistering, swelling, redness, heat, hemorrhaging (bleeding), and gangrene.

• *First aid measures.* First aid measures are required for all stages of immersion syndrome injury. Rewarm the injured part gradually by exposing it to warm air. Protect it from trauma and secondary infections. Dry, loose clothing or several layers of warm coverings are preferable to extreme heat. Under no circumstances should the injured part be exposed to an open fire. Elevate the injured part to relieve the swelling. Transport the casualty to an MTF as soon as possible. When the part is rewarmed, the casualty often feels a burning sensation and pain. Symptoms may persist for days or weeks even after rewarming.

NOTE

When providing first aid for immersion foot and trench foot—
DO NOT massage the injured part. **DO NOT** moisten the skin.
DO NOT apply heat or ice.

(3) *Frostbite*. Frostbite is the injury of tissue caused from exposure to cold, usually below 32°F depending on the windchill factor, duration of exposure, and adequacy of protection. Individuals with a history of cold injury are likely to suffer an additional cold injury. The body parts most easily frostbitten are the cheeks, nose, ears, chin, forehead, wrists, hands, and feet. Frostbite may involve only the skin (superficial), or it may extend to a depth below the skin (deep). Deep frostbite is very serious and requires prompt first aid to avoid or to minimize the loss of parts or all of the fingers, toes, hands, or feet.

- *Signs and symptoms*.
  - Loss of sensation (numb feeling) in any part of the body.
  - Sudden blanching (whitening) of the skin of the affected part, followed by a momentary tingling sensation.
  - Redness of skin in light-skinned service members; grayish coloring in dark-skinned service members.
  - Blisters.
  - Swelling or tender areas.
  - Loss of previous sensation of pain in affected area.
  - Pale, yellowish, waxy-looking skin.
  - Frozen tissue that feels solid (or wooden) to the touch.

CAUTION

Deep frostbite is a very serious injury and requires immediate first aid and subsequent medical treatment to avoid or minimize loss of body parts.

- *First aid measures.*

  - *Face, ears, and nose.* Cover the casualty's affected area with his and/or your bare hands until sensation and color return.

  - *Hands.* Open the casualty's field jacket and shirt. (In a chemical environment, do not loosen or remove the clothing and protective overgarments.) Place the affected hands under the casualty's armpits. Close the field jacket and shirt to prevent additional exposure.

  - *Feet.* Remove the casualty's boots and socks if he does not need to walk any further to receive additional treatment. (Thawing the casualty's feet and forcing him to walk on them will cause additional pain and injury.) Place the affected feet under clothing and against the body of another service member.

WARNING

DO NOT attempt to thaw the casualty's feet or other frozen areas if he will be required to walk or travel to an MTF for additional medical treatment. The possibility of additional injury from walking is less when the feet are frozen than when they are thawed. (However, if possible avoid walking.) Thawing in the field increases the possibilities of infection, gangrene, or other injury.

NOTE

Thawing may occur spontaneously during transportation to the MTF; this cannot be avoided since the body in general must be kept warm.

In all of the above areas, ensure that the casualty is kept warm and that he is covered (to avoid further injury). Seek medical treatment as soon as possible. Reassure the casualty, protect the affected area from further injury by covering it lightly with a blanket or any dry clothing, and seek shelter out of the wind. Remove or loosen constricting clothing (except in a contaminated environment) and increase insulation. Ensure the casualty exercises as much as possible, avoiding trauma to the injured part, and is prepared for pain when thawing occurs. Protect the frostbitten part from additional injury. **DO NOT**—

- Rub the injured part with snow or apply cold water soaks.

- Warm the part by massage or exposure to open fire because the frozen part may be burned due to the lack of feeling.

- Use ointments or other salves.

- Manipulate the part in any way to increase circulation.

- Use alcohol or tobacco because this reduces the body's resistance to cold.

NOTE

> Remember, when freezing extends to a depth below the skin, it is a much more serious injury. Extra care is required to reduce or avoid the chances of losing all or part of the toes or feet. This also applies to the fingers and hands.

(4) *Snow blindness.* Snow blindness is the effect that glare from an ice field or snowfield has on the eyes. It is more likely to occur in hazy, cloudy weather than when the sun is shining. Glare from the sun will cause an individual to instinctively protect his eyes. However, in cloudy weather, he may be overconfident and expose his eyes longer than when the threat is more obvious. He may also neglect precautions such as the use of protective eyewear. Waiting until discomfort (pain) is felt before using protective eyewear is dangerous because a deep burn of the eyes may already have occurred.

- *Signs and symptoms.* Symptoms of snow blindness are a sensation of grit in the eyes with pain in and over the eyes, made worse by moving the eyeball. Other signs and symptoms are watering, redness, headache, and increased pain on exposure to light.

- *First aid measures.* First aid measures consist of blindfolding or covering the eyes with a dark cloth which stops painful eye movement. Complete rest is desirable. If further exposure to light is not preventable, the eyes should be protected with dark bandages or the darkest glasses available. Once unprotected exposure to sunlight stops, the condition usually heals in a few days without permanent damage. The casualty should be evacuated to the nearest MTF.

(5) *Dehydration.* Dehydration occurs when the body loses too much fluid. A certain amount of body fluid is lost through normal body processes. A normal daily intake of liquids replaces these losses. When individuals are engaged in any strenuous exercises or activities, fluid is lost

through sweating and this loss creates an imbalance of fluids in the body, and if not matched by rehydration it can contribute to dehydration. The danger of dehydration is as prevalent in cold regions as it is in hot regions. In hot weather, the individual is aware of his body losing fluids through sweat. In cold weather, however, it is extremely difficult to realize that this condition exists since sweating is not as apparent as in a hot environment. The danger of dehydration in cold weather operations is a serious problem. In cold climates, sweat evaporates so rapidly or is absorbed so thoroughly by layers of heavy clothing that it is rarely visible on the skin. Dehydration also occurs during cold weather operations because drinking is inconvenient. Dehydration will weaken or incapacitate a casualty for a few hours, or sometimes, several days. Because rest is an important part of the recovery process, casualties must take care that limited movement during their recuperative period does not enhance the risk of becoming a cold injury casualty.

- *Signs and symptoms.* The symptoms of cold weather dehydration are similar to those encountered in heat exhaustion. The mouth, tongue, and throat become parched and dry, and swallowing becomes difficult. The casualty may have nausea (with or without vomiting) along with extreme dizziness and fainting. The casualty may also feel generally tired and weak and may experience muscle cramps. Focusing the eyes may also become difficult.

- *First aid measures.* The casualty should be kept warm and his clothes should be loosened (if not in a chemical environment) to allow proper circulation. Shelter from wind and cold must be provided. Fluid replacement should begin immediately and the service member transported to an MTF as soon as possible.

(6) *Hypothermia (general cooling).* When exposed to prolonged cold weather a service member may become both mentally and physically numb, thus neglecting essential tasks or requiring more time and effort to achieve them. Under some conditions (particularly cold water immersion), even a service member in excellent physical condition may die in a matter of minutes. The destructive influence of cold on the body is called *hypothermia.* This means bodies lose heat faster than they can produce it. Hypothermia can occur from exposure to temperatures either above or below freezing, especially from immersion in cold water, wet-cold conditions, or from the effect of wind. Physical exhaustion and insufficient food intake may also increase the risk of hypothermia. General cooling of the entire body to a temperature below 95°F is caused by continued exposure to low or rapidly dropping temperatures, cold moisture, snow, or ice. Fatigue, poor physical condition, dehydration, faulty blood circulation, alcohol or other drug use, trauma, and immersion can cause hypothermia. Remember, cold

may affect the body systems slowly and almost without notice. Service members exposed to low temperatures for extended periods may suffer ill effects even if they are well protected by clothing.

- *Signs and symptoms*. As the body cools, there are several stages of progressive discomfort and impairment. A sign that is noticed immediately is shivering. Shivering is an attempt by the body to generate heat. The pulse is faint or very difficult to detect. People with temperatures around 90°F may be drowsy and mentally slow. Their ability to move may be hampered, stiff, and uncoordinated, but they may be able to function minimally. Their speech may be slurred. As the body temperature drops further, shock becomes evident as the person's eyes assume a glassy state, breathing becomes slow and shallow, and the pulse becomes weaker or absent. The person becomes very stiff and uncoordinated. Unconsciousness may follow quickly. As the body temperature drops even lower, the extremities freeze, and a deep (or core) body temperature (below 85°F) increases the risk of irregular heart action. This irregular heart action or heart standstill can result in sudden death.

- *First aid measures*. Except in cases of the most severe hypothermia (marked by coma or unconsciousness and a weak pulse), first aid measures for hypothermia are directed towards protecting the casualty from further loss of body heat. For the casualty who is conscious, first aid measures are directed at rewarming the body evenly and without delay. Provide heat by using a hot water bottle or field expedient or another service member's body heat.

CAUTION

DO NOT expose the casualty to an open fire, as he may become burned.

NOTE

When using a hot water bottle or field expedient (canteen filled with warm water), the bottle or canteen must be wrapped in cloth prior to placing it next to the casualty. This will reduce the chance of burning the casualty's skin.

Always call or send for help as soon as possible and protect the casualty immediately with dry clothing or a sleeping bag. Then, move him to a warm place. Evaluate other injuries and provide first aid as required. First aid measures can be performed while the casualty is waiting transportation or while he is en route. In the case of an accidental breakthrough into ice water, or other hypothermic accident, strip the casualty of wet clothing immediately and bundle him into a sleeping bag. Rescue breathing should be started at once if the casualty's breathing has stopped or is irregular or shallow. Warm liquids (**NOT HOT**) may be given gradually if the casualty is conscious. **DO NOT** force liquids on an unconscious or semiconscious casualty because he may choke. The casualty should be transported on a litter because the exertion of walking may aggravate circulation problems. Medical personnel should immediately treat any hypothermia casualty. Hypothermia is life threatening until normal body temperature has been restored. The first aid measures for a casualty with severe hypothermia are based upon the following principles: attempt to avoid further heat loss, handle the casualty gently, and transport the casualty as soon as possible to the nearest MTF. If at all possible, the casualty should be evacuated by medical personnel.

WARNING

Rewarming a severely hypothermic casualty is extremely dangerous in the field due to the possibility of such complications as rewarming, shock and disturbances in the rhythm of the heartbeat. These conditions require treatment by medical personnel.

NOTE

Resuscitation of casualties with hypothermic complications is difficult if not impossible to do outside of an MTF setting.

CAUTION

The casualty is unable to generate his own body heat. Therefore, merely placing him in a blanket or sleeping bag is not sufficient.

e. *Table*. See Table 5-2 for further information.

Table 5-2. Injuries Caused by Cold and Wet Conditions

INJURIES	SIGNS/SYMPTOMS	FIRST AID
CHILBLAIN	RED SWOLLEN, HOT, TENDER, ITCHING SKIN. CONTINUED EXPOSURE MAY LEAD TO INFECTED (ULCERATED OR BLEEDING) SKIN LESIONS.	1. AREA USUALLY RESPONDS TO LOCALLY APPLIED REWARMING (BODY HEAT). 2. DO NOT RUB OR MASSAGE AREA. 3. SEEK MEDICAL AID.
IMMERSION SYNDROME (IMMERSION FOOT/TRENCH FOOT)	AFFECTED PARTS ARE COLD, NUMB, AND PAINLESS. PARTS MAY THEN BE HOT, WITH BURNING AND SHOOTING PAINS. ADVANCED STAGE: SKIN PALE WITH BLUISH CAST; PULSE DECREASES; BLISTERING, SWELLING, HEAT, HEMORRHAGING, AND GANGRENE MAY FOLLOW.	1. GRADUAL REWARMING BY EXPOSURE TO WARM AIR. 2. DO NOT MASSAGE OR MOISTEN SKIN. 3. PROTECT AFFECTED PARTS FROM TRAUMA. 4. DRY FEET THOROUGHLY, AVOID WALKING. 5. SEEK MEDICAL AID.
FROSTBITE	LOSS OF SENSATION (NUMB FEELING) IN ANY PART OF THE BODY. SUDDEN BLANCHING (WHITENING) OF THE SKIN OF THE AFFECTED PART, FOLLOWED BY A MOMENTARY TINGLING SENSATION. REDNESS OF SKIN IN LIGHT-SKINNED SERVICE MEMBERS; GRAYISH COLORING IN DARK-SKINNED SERVICE MEMBERS. BLISTERS. SWELLING OR TENDER AREAS. LOSS OF PREVIOUS SENSATION OF PAIN IN THE AFFECTED AREA. PALE YELLOWISH, WAXY-LOOKING SKIN. FROZEN TISSUE THAT FEELS SOLID (WOODEN) TO THE TOUCH.	1. WARM THE AREA AT THE FIRST SIGN OF FROSTBITE, USING FIRM, STEADY PRESSURE OF THE HAND, UNDERARM, OR ABDOMEN. 2. FACE, EARS, NOSE: COVER AREA WITH HANDS (CASUALTY'S OWN OR BUDDY'S). 3. HANDS: OPEN FIELD JACKET AND PLACE CASUALTY'S HANDS AGAINST HIS BODY, THEN CLOSE THE JACKET TO PREVENT HEAT LOSS. 4. FEET: REMOVE THE CASUALTY'S BOOTS AND SOCKS AND PLACE HIS FEET AGAINST THE BODY OF ANOTHER SERVICE MEMBER.

Table 5-2. Injuries Caused by Cold and Wet Conditions (Continued)

INJURIES	SIGNS/SYMPTOMS	FIRST AID
		5. WARNING: DO NOT ATTEMPT TO THAW THE CASUALTY'S FEET OR OTHER FROZEN AREAS IF HE WILL BE REQUIRED TO WALK OR TRAVEL TO AN MTF FOR ADDITIONAL TREATMENT. THE POSSIBILITY OF INJURY FROM WALKING IS LESS WHEN THE FEET ARE FROZEN THAN WHEN THEY HAVE BEEN THAWED. (HOWEVER, IF POSSIBLE AVOID WALKING.) THAWING IN THE FIELD INCREASES THE POSSIBILITY OF INFECTION, GANGRENE, OR INJURY. 6. LOOSEN OR REMOVE CONSTRICTING CLOTHING AND REMOVE ANY JEWELRY. 7. INCREASE INSULATION (COVER WITH BLANKET OR OTHER DRY MATERIAL). ENSURE CASUALTY EXERCISES AS MUCH AS POSSIBLE, AVOIDING TRAUMA TO INJURED PART.
SNOW BLINDNESS	EYES MAY FEEL SCRATCHY. WATERING, REDNESS, HEADACHE, AND INCREASED PAIN WITH EXPOSURE TO LIGHT CAN OCCUR.	1. COVER THE EYES WITH A DARK CLOTH. 2. SEEK MEDICAL AID.
DEHYDRATION	SIMILAR TO HEAT EXHAUSTION (REFER TO TABLE 5-1).	1. KEEP WARM. 2. CASUALTY NEEDS FLUID REPLACEMENT, REST, AND PROMPT MEDICAL AID.
HYPOTHERMIA	CASUALTY IS COLD. SHIVERING. CORE TEMPERATURE IS LOW. CONSCIOUSNESS MAY BE	**MILD HYPOTHERMIA** 1. REWARM BODY

Table 5-2. Injuries Caused by Cold and Wet Conditions (Continued)

INJURIES	SIGNS/SYMPTOMS	FIRST AID
	ALTERED. UNCOORDINATED MOVEMENTS MAY OCCUR. SHOCK AND COMA MAY RESULT AS BODY TEMPERATURE DROPS.	EVENLY AND WITHOUT DELAY. (NEED TO PROVIDE HEAT SOURCE; CASUALTY'S BODY UNABLE TO GENERATE HEAT.) 2. KEEP DRY, PROTECT FROM THE ELEMENTS. 3. WARM (NOT HOT) LIQUIDS MAY BE GIVEN GRADUALLY (TO CONSCIOUS CASUALTIES ONLY). 4. BE PREPARED TO START BASIC LIFE SUPPORT MEASURES FOR THE CASUALTY. 5. SEEK MEDICAL TREATMENT IMMEDIATELY. **SEVERE HYPOTHERMIA** 1. STABILIZE THE TEMPERATURE. 2. ATTEMPT TO AVOID FURTHER HEAT LOSS. 3. HANDLE THE CASUALTY GENTLY. 4. EVACUATE TO THE NEAREST MTF AS SOON AS POSSIBLE. **5. WARNING: HYPOTHERMIA IS A MEDICAL EMERGENCY. PROMPT MEDICAL ATTENTION IS NECESSARY.**

CHAPTER 6

FIRST AID FOR BITES AND STINGS

6-1. General

Snakebites, insect bites, or stings can cause intense pain and/or swelling. If not treated promptly and correctly, they can cause serious illness or death. The severity of a snakebite depends upon: whether the snake is poisonous or nonpoisonous, the type of snake, the location of the bite, and the amount of venom injected. Bites from humans and other animals, such as dogs, cats, bats, raccoons, and rats, can cause severe bruises and infection and tears or lacerations of tissue. Awareness of the potential sources of injuries can reduce or prevent them from occurring. Knowledge and prompt application of first-aid measures can lessen the severity of injuries from bites and stings and keep the service member from becoming a serious casualty.

6-2. Types of Snakes

a. Nonpoisonous Snakes. There are approximately 130 different varieties of nonpoisonous snakes in the United States. They have oval-shaped heads and round eyes. Unlike poisonous snakes, discussed below, nonpoisonous snakes do not have fangs with which to inject venom. Figure 6-1 depicts the characteristics of a nonpoisonous snake.

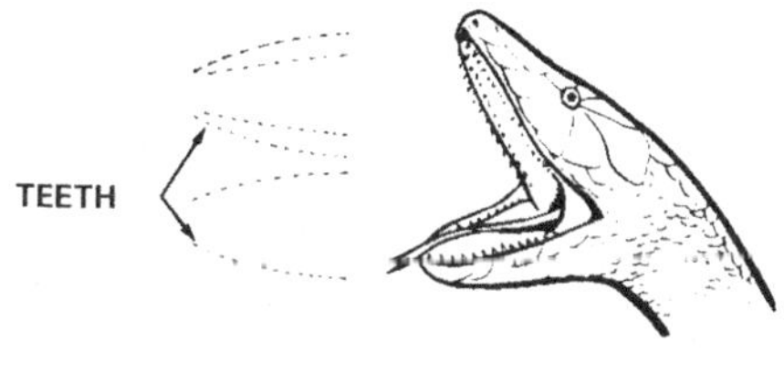

Figure 6-1. Characteristics of nonpoisonous snake.

b. Poisonous Snakes. Poisonous snakes are found throughout the world, primarily in tropical to moderate climates. Within the United States, there are four kinds: rattlesnakes, copperheads, water moccasins (cottonmouth), and coral snakes. Poisonous snakes in other parts of the world include sea snakes, the fer-de-lance, the bushmaster, and the tropical rattlesnake in tropical Central America; the Malayan pit viper in the tropical Far East; the cobra in Africa and Asia; the mamba (or black mamba) in central and southern Africa; and the krait in India and Southeast Asia. Refer to Figure 6-2 for characteristics of a poisonous pit viper.

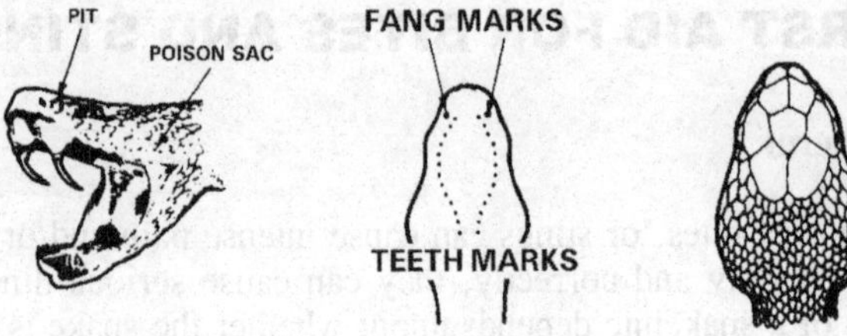

Figure 6-2. Characteristics of poisonous pit viper.

c. Pit Vipers (Poisonous). Figure 6-3 depicts a variety of poisonous snakes.

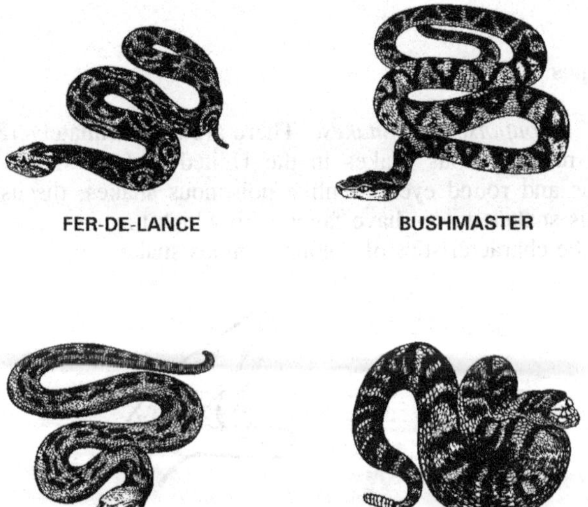

Figure 6-3. Poisonous snakes.

(1) Rattlesnakes, bushmasters, copperheads, fer-de-lance, Malayan pit vipers, and water moccasins (cottonmouth) are called pit vipers because of the small, deep pits between the nostrils and eyes on each side of the head (Figure 6-2). In addition to their long, hollow fangs, these snakes have other identifying features: thick bodies, slit-like pupils of the eyes, and flat, almost triangular-shaped heads. Color markings and other identifying characteristics, such as rattles or a noticeable white interior of the mouth (cottonmouth), also help distinguish these poisonous snakes. Further

identification is provided by examining the bite pattern of the wound for signs of fang entry. Occasionally there will be only one fang mark, as in the case of a bite on a finger or toe where there is no room for both fangs, or when the snake has broken off a fang.

(2) The casualty's condition provides the best information about the seriousness of the situation, or how much time has passed since the bite occurred. Pit viper bites are characterized by severe burning pain. Discoloration and swelling around the fang marks usually begins within 5 to 10 minutes after the bite. If only minimal swelling occurs within 30 minutes, the bite will almost certainly have been from a nonpoisonous snake or possibly from a poisonous snake which did not inject venom. The venom destroys blood cells, causing a general discoloration of the skin. Blisters and numbness in the affected area follow this reaction. Other signs, which can occur, are weakness, rapid pulse, nausea, shortness of breath, vomiting, and shock.

d. Corals, Cobras, Kraits, and Mambas. Corals (Figure 6-4), cobras (Figure 6-5), kraits, and mambas all belong to the same group even though they are found in different parts of the world. All four inject their venom through short, grooved fangs, leaving a characteristic bite pattern.

Figure 6-4. Coral snake.

(1) The small coral snake, found in the Southeastern US, is brightly colored with bands of red, yellow (or almost white), and black completely encircling the body. Other nonpoisonous snakes have the same coloring, but on the coral snake found in the US, the red ring always touches the yellow ring. To know the difference between a harmless snake and the coral snake found in the United States, remember the following:

"Red on yellow will kill a fellow,
Red on black, venom will lack."

Figure 6-5. Cobra snake.

(2) The venom of corals, cobras, kraits, and mambas produces symptoms different from those of pit vipers. Because there is only minimal pain and swelling, many people believe that the bite is not serious. Delayed reactions in the nervous system normally occur between 1 to 7 hours after the bite. Symptoms include blurred vision, drooping eyelids, slurred speech, drowsiness, and increased salivation and sweating. Nausea, vomiting, shock, respiratory difficulty, paralysis, convulsions, and coma will usually develop if the bite is not treated promptly.

e. Sea Snakes. Sea snakes (Figure 6-6) are found in the warm water areas of the Pacific and Indian oceans, along the coasts, and at the mouths of some larger rivers. Their venom is **VERY** poisonous, but their fangs are only 1/4 inch long. The first aid outlined for land snakes also applies to sea snakes.

Figure 6-6. Sea snake.

6-3. Snakebites

a. Poisonous snakes **DO NOT** always inject venom when they bite or strike a person. However, all snakes may carry tetanus (lockjaw); anyone bitten by a snake, whether poisonous or nonpoisonous, should immediately seek medical attention.

- Poison is injected from the venom sacs through grooved or hollow fangs. Depending on the species, these fangs are either long or short. Pit vipers have long hollow fangs. These fangs are folded against the roof of the mouth and extend when the snake strikes. This allows them to strike quickly and then withdraw. Cobras, coral snakes, kraits, mambas, and sea snakes have short, grooved fangs. These snakes are less effective in their attempts to bite, since they must chew after striking to inject enough venom (poison) to be effective. Figure 6-7 depicts the characteristics of a poisonous snakebite.

- In the event you are bitten, attempt to identify and/or kill the snake. Take it to medical personnel for inspection/identification. This provides valuable information to medical personnel who deal with snakebites. **TREAT ALL SNAKEBITES AS POISONOUS.**

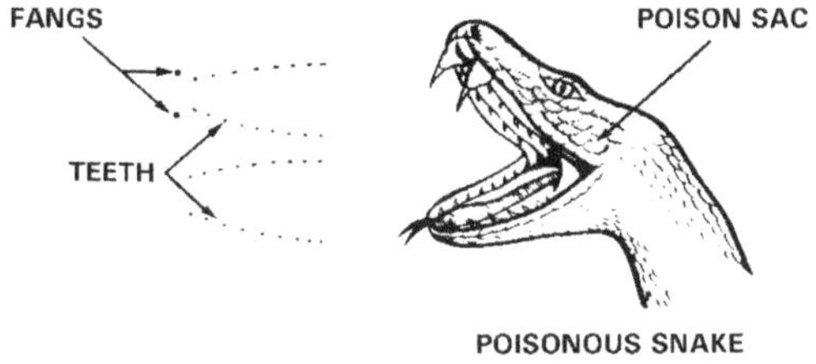

Figure 6-7. Characteristics of poisonous snakebite.

b. The venoms of different snakes cause different effects. Pit viper venom (hemotoxin [blood toxin]) destroys tissue and blood cells. Cobras, adders, and coral snakes inject powerful venom (neurotoxin [nerve toxin]) which affect the central nervous system, causing respiratory paralysis. Water moccasins and sea snakes have venom that is both hemotoxic and neurotoxic.

c. The identification of poisonous snakes is very important since medical treatment will be different for each type of venom. *Unless it can be positively identified, the snake should be killed and saved.* When this is not possible or when doing so is a serious threat to others, identification may sometimes be difficult since many venomous snakes resemble harmless varieties. When dealing with snakebite problems in foreign countries, seek advice, professional or otherwise, which may help identify species in the particular area of operations.

d. Get the casualty to an MTF as soon as possible and with minimum movement. Until evacuation or treatment is possible, have the casualty lie quietly and not move any more than necessary. If the casualty has been bitten on an extremity, **DO NOT** elevate the limb; keep the extremity level with the body. Keep the casualty comfortable and reassure him. If the casualty is alone when bitten, he should go to the medical facility himself rather than wait for someone to find him. Unless the snake has been positively identified, attempt to kill it and send it with the casualty. Be sure that retrieving the snake does not endanger anyone or delay transporting the casualty.

(1) If the bite is on an arm or leg, place a constricting band (narrow cravat [swathe], or narrow gauze bandage) one to two fingerbreadths above and below the bite (Figure 6-8). If the bite is on the hand or foot, place a single band above the wrist or ankle. The band should be tight enough to stop the flow of blood near the skin, but not tight enough to interfere with circulation. In other words, it should not have a tourniquet-like affect. If no swelling is seen, place the bands about 1 inch from either side of the bite. If swelling is present, put the bands on the unswollen part at the edge of the swelling. If the swelling extends beyond the band, move the band to the new edge of the swelling. (If possible, leave the old band on, place a new one at the new edge of the swelling, and then remove and save the old one in case the process has to be repeated.)

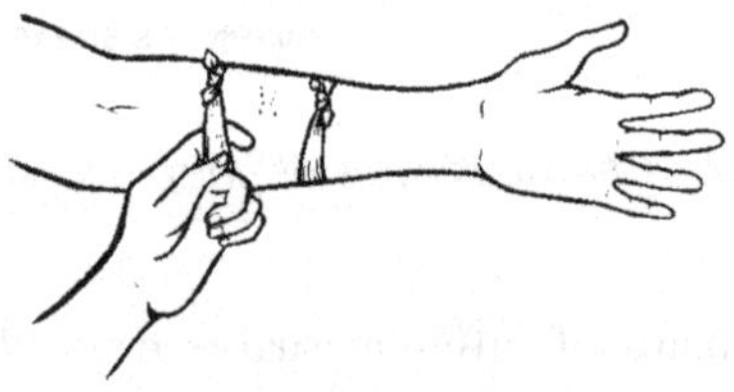

Figure 6-8. Constricting band.

CAUTION

DO NOT attempt to cut open the bite nor suck out the venom. If the venom should seep through any damaged or lacerated tissues in your mouth, you could immediately lose consciousness or even die.

(2) If the bite is located on an arm or leg, immobilize it at a level below the heart. **DO NOT** elevate an arm or leg even with or above the level of the heart.

CAUTION

When a splint is used to immobilize the arm or leg, take **EXTREME** care to ensure the splinting is done properly and does not bind. Watch it closely and adjust it if any changes in swelling occur.

(3) When possible, clean the area of the bite with soap and water. **DO NOT** use ointments of any kind.

(4) **NEVER** give the casualty food, alcohol, stimulants (coffee or tea), drugs, or tobacco.

(5) Remove rings, watches, or other jewelry from the affected limb.

6-4. Human or Animal Bites

Human or other land animal bites may cause lacerations or bruises. In addition to damaging tissue, bites always present the possibility of infection.

a. Human Bites. Human bites that break the skin may become seriously infected since the mouth is heavily contaminated with bacteria. Medical personnel **MUST** treat all human bites.

b. Animal Bites. Land animal bites can result in both infection and disease. Tetanus, rabies, and various types of fevers can follow an untreated animal bite. Because of these possible complications, the animal causing the bite should, if possible, be captured or killed (without damaging its head) so that it can be tested for disease.

c. First Aid.

(1) Cleanse the wound thoroughly with soap.

(2) Flush it well with water.

(3) Cover it with a sterile dressing.

(4) Immobilize the injured arm or leg, if appropriate.

(5) Transport the casualty immediately to an MTF.

NOTE

If unable to capture or kill the animal, provide medical personnel with any information that will help identify it.

6-5. Marine (Sea) Animals

With the exception of sharks and barracuda, most marine animals will not deliberately attack. The most frequent injuries from marine animals are wounds by biting, stinging, or puncturing. Wounds inflicted by marine animals can be very painful, but are rarely fatal.

a. Sharks, Barracuda, and Alligators. Wounds from these marine animals can involve major trauma as a result of bites and lacerations. Bites from large marine animals are potentially the most life threatening of all injuries from marine animals. Major wounds from these animals can be treated by controlling the bleeding, preventing shock, giving basic life support, splinting the injury, and by securing prompt medical aid.

b. Turtles, Moray Eels, and Corals. These animals normally inflict minor wounds. Treat by cleansing the wound(s) thoroughly and by splinting if necessary.

c. Jellyfish, Portuguese Man-of-War, Anemones, and Others. This group of marine animals inflict injury by means of stinging cells in their tentacles. Contact with the tentacles produces burning pain with a rash and small hemorrhages on the skin. Shock, muscular cramping, nausea, vomiting, and respiratory distress may also occur. Gently remove the clinging tentacles with a towel and wash or treat the area. Use diluted ammonia or alcohol, meat tenderizer, and talcum powder. If symptoms become severe or persist, seek medical assistance.

d. Spiny Fish, Urchins, Stingrays, and Cone Shells. These animals inject their venom by puncturing the skin with their spines. General signs and symptoms include swelling, nausea, vomiting, generalized cramps, diarrhea, muscular paralysis, and shock. Deaths are rare. Treatment consists of soaking the wounds in hot water (when available) for 30 to 60 minutes. This inactivates the heat sensitive toxin. In addition, further first aid measures (controlling bleeding, applying a dressing, and so forth) should be carried out as necessary.

CAUTION

Be careful not to scald the casualty with water that is too hot because the pain of the wound will mask the normal reaction to heat.

6-6. Insect (Arthropod) Bites and Stings

An insect bite or sting can cause great pain, allergic reaction, inflammation, and infection. If not treated correctly, some bites/stings may cause serious illness or even death. When an allergic reaction is not involved, first aid is a simple process. In any case, medical personnel should examine the casualty at the earliest possible time. It is important to properly identify the spider, bee, or creature that caused the bite/sting, especially in cases of allergic reaction.

a. Types of Insects. The insects found throughout the world that can produce a bite or sting are too numerous to mention in detail. Commonly encountered stinging or biting insects include brown recluse spiders (Figure 6-9), black widow spiders (Figure 6-10), tarantulas (Figure 6-11), scorpions (Figure 6-12), urticating caterpillars, bees, wasps, centipedes, conenose beetles (kissing bugs), ants, and wheel bugs. Upon being reassigned, especially to overseas areas, take the time to become acquainted with the types of insects to avoid.

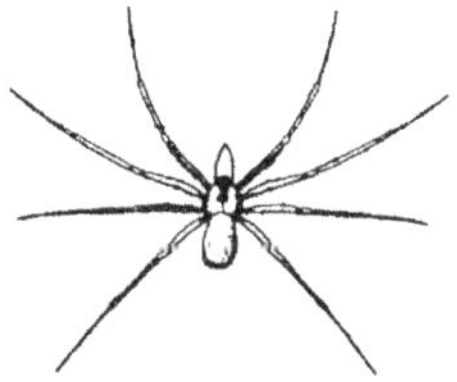

Figure 6-9. Brown recluse spider.

Figure 6-10. Black widow spider.

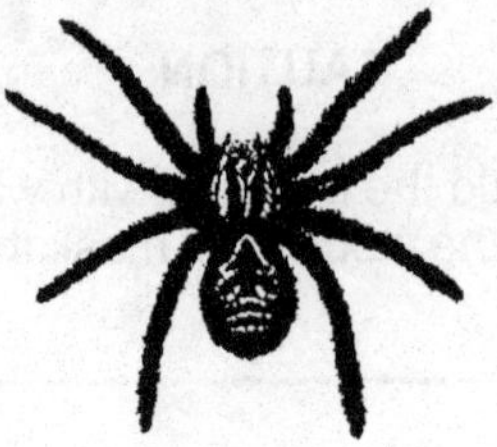

Figure 6-11. Tarantula.

Figure 6-12. Scorpion.

b. Signs and Symptoms. Discussed in paragraphs (1) and (2) below are the most common effects of insect bites/stings. They can occur alone or in combination with the others.

(1) *Less serious.* Commonly seen signs/symptoms are pain, irritation, swelling, heat, redness, and itching. Hives or wheals (raised areas of the skin that itch) may occur. These are the least severe of the allergic reactions that commonly occur from insect bites/stings. They are usually dangerous only if they affect the air passages (mouth, throat, nose, and so forth), which could interfere with breathing. The bites/stings of bees, wasps, ants, mosquitoes, fleas, and ticks are usually not serious and normally produce mild and localized symptoms. A tarantula's bite is usually no worse than that of a bee sting. Scorpions are rare and their stings (except for a specific species found only in the Southwest desert) are painful but usually not dangerous.

(2) *Serious.* Emergency allergic or hypersensitive reactions sometimes result from the stings of bees, wasps, and ants. Many people are allergic to the venom of these particular insects. Bites or stings from these insects may produce more serious reactions, to include generalized itching and hives, weakness, anxiety, headache, breathing difficulties, nausea, vomiting, and diarrhea. Very serious allergic reactions (called *anaphylactic shock*) can lead to complete collapse, shock, and even death. Spider bites (particularly from the black widow and brown recluse spiders) can also be

serious. Venom from the black widow spider affects the nervous system. This venom can cause muscle cramps, a rigid, nontender abdomen, breathing difficulties, sweating, nausea, and vomiting. The brown recluse spider generally produces local rather than system-wide problems; however, local tissue damage around the bite can be severe and can lead to an ulcer and even gangrene.

c. *First Aid.* There are certain principles that apply regardless of what caused the bite/sting. Some of these are—

- If there is a stinger present (for example, from a bee), remove the stinger by scraping the skin's surface with a fingernail or knife. **DO NOT** squeeze the sac attached to the stinger because it may inject more venom.

- Wash the area of the bite/sting with soap and water (alcohol or an antiseptic may also be used) to help reduce the chances of an infection and remove traces of venom.

- Remove jewelry from bitten extremities because swelling may occur.

- In most cases of insect bites the reaction will be mild and localized; use ice or cold compresses (if available) on the site of the bite/sting. This will help reduce swelling, ease the pain, and slow the absorption of venom. Meat tenderizer (to neutralize the venom) or calamine lotion (to reduce itching) may be applied locally. If necessary, seek medical assistance.

- In more serious reactions (severe and rapid swelling, allergic symptoms, and so forth) treat the bite/sting like you would treat a snakebite; that is, apply constricting bands above and below the site.

- Be prepared to perform basic life-support measures, such as rescue breathing.

- Reassure the casualty and keep him calm.

- In serious reactions, attempt to capture the insect for positive identification; however, be careful not to become a casualty yourself.

- If the reaction to the bite/sting appears serious, seek medical assistance.

WARNING

Insect bites/stings may cause *anaphylactic shock* (a shock caused by a severe allergic reaction). This is a life-threatening event and a TRUE MEDICAL EMERGENCY. Be prepared to perform the basic life-support measures and to immediately transport the casualty to an MTF.

NOTE

Be aware that some allergic or hypersensitive individuals may carry identification or emergency insect bite treatment kits. If the casualty is having an allergic reaction and has such a kit, administer the medication in the kit according to the instructions which accompany the kit.

d. Supplemental Information. For additional information concerning biting insects, see FM 21-10.

6-7. First Aid for Bites and Stings

See the table below for information on bites and stings.

Table 6-1. First Aid Measures for Bites and Stings

TYPES	FIRST AID MEASURES
SNAKEBITE	1. MOVE CASUALTY AWAY FROM THE SNAKE. 2. REMOVE JEWELRY FROM THE AFFECTED AREA, IF APPLICABLE. 3. REASSURE CASUALTY AND KEEP HIM QUIET. 4. APPLY CONSTRICTING BAND, 1-2 FINGERBREADTHS FROM THE BITE. YOU SHOULD BE ABLE TO INSERT A FINGER BETWEEN THE BAND AND THE SKIN. *a. ARM OR LEG BITE.* PLACE ONE BAND ABOVE AND ONE BAND BELOW THE BITE SITE. *b. HAND OR FOOT BITE.* PLACE ONE BAND ABOVE THE WRIST OR ANKLE.

Table 6-1. First Aid Measures for Bites and Stings

TYPES	FIRST AID MEASURES
	5. IMMOBILIZE THE AFFECTED PART IN A POSITION BELOW THE LEVEL OF THE HEART. 6. KILL THE SNAKE (IF POSSIBLE, WITHOUT DAMAGING ITS HEAD OR ENDANGERING YOURSELF) AND SEND IT WITH THE CASUALTY. 7. SEEK MEDICAL ASSISTANCE IMMEDIATELY.
BROWN RECLUSE SPIDER OR BLACK WIDOW SPIDER BITE	1. KEEP CASUALTY QUIET. 2. REMOVE ALL JEWELRY FROM AFFECTED PART, IF APPLICABLE. 3. WASH THE AREA. 4. APPLY ICE OR FREEZE PACK, IF AVAILABLE. 5. SEEK MEDICAL ASSISTANCE.
TARANTULA BITE OR SCORPION STING OR ANT BITE	1. WASH THE AREA. 2. REMOVE ALL JEWELRY FROM AFFECTED PART, IF APPLICABLE. 3. APPLY ICE OR FREEZE PACK, IF AVAILABLE. 4. APPLY BAKING SODA, CALAMINE LOTION, OR MEAT TENDERIZER (IF AVAILABLE) TO BITE SITE TO RELIEVE PAIN AND ITCHING. 5. IF THE SITE OF THE BITE IS ON THE FACE, NECK (POSSIBLE AIRWAY PROBLEMS), OR GENITAL AREA, OR IF LOCAL REACTION SEEMS SEVERE, OR IF THE STING IS BY THE DANGEROUS TYPE OF SCORPION FOUND IN THE SOUTHWEST UNITED STATES DESERT, KEEP THE CASUALTY AS QUIET AS POSSIBLE. SEEK MEDICAL ASSISTANCE.
BEE STING	1. IF THE STINGER IS PRESENT, REMOVE BY SCRAPING WITH A KNIFE OR FINGERNAIL. DO NOT SQUEEZE VENOM SAC ON STINGER; MORE VENOM MAY BE INJECTED. 2. REMOVE ALL JEWELRY FROM AFFECTED PART, IF APPLICABLE.

Table 6-1. First Aid Measures for Bites and Stings

TYPES	FIRST AID MEASURES
	3. WASH THE AREA. 4. APPLY ICE OR FREEZE PACK, IF AVAILABLE. 5. IF ALLERGIC SIGNS OR SYMPTOMS APPEAR, BE PREPARED TO PERFORM BASIC LIFE SUPPORT MEASURES. SEEK IMMEDIATE MEDICAL ASSISTANCE.

CHAPTER 7

FIRST AID IN A NUCLEAR, BIOLOGICAL, AND CHEMICAL ENVIRONMENT

7-1. General

American forces have not been exposed to NBC weapons/agents on the battlefield since World War I. In future conflicts and wars we can expect the use of such agents. Nuclear, biological, and chemical weapons will rapidly degrade unit effectiveness by forcing troops to wear protective clothing and by creating confusion and fear. Through training in protective procedures and first aid, units can maintain their effectiveness on the integrated battlefield.

7-2. First Aid Materials

You may be issued the following materials to protect, decontaminate, and use as first aid for NBC exposure. You must know how to use the items; some items are described in *a* through *d* below. It is equally important that you know when to use them.

a. Nerve Agent Pyridostigmine Pretreatment (*NAPP*). You may be issued a blister pack of pretreatment tablets when your commander directs. The NAPP is a pretreatment; it is not an antidote. It improves the effectiveness of the nerve agent antidote. When ordered to take the pretreatment you must take one tablet every 8 hours, mission permitting. This must be taken prior to exposure to nerve agents, since it may take several hours to develop adequate blood levels.

NOTE

Commanders must follow investigational new drug protocols for use of the NAPP.

b. M291 Skin Decontaminating Kit. The M291 Skin Decontaminating Kit (Figure 7-1) contains six packets of XE-555 decontaminant resin.

WARNING

For external use only. May be slightly irritating to the eyes. Keep decontaminating powder out of eyes. Use water to wash toxic agent out of eyes.

c. Nerve Agent Antidote Kit, MARK I. Each service member is issued three MARK Is for use in first aid for nerve agent poisoning (Figure 7-2 and paragraph 7-6).

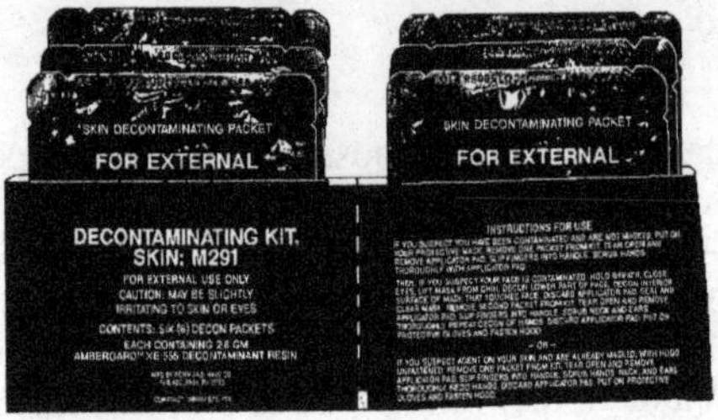

Figure 7-1. M291 Skin Decontamination Kit.

d. Antidote Treatment, Nerve Agent, Autoinjector. A new nerve agent antidote injection device, Antidote Treatment, Nerve Agent, Autoinjector (ATNAA) is currently under development that will replace the MARK I. The ATNAA is a multichambered device with the atropine and pralidoxime chloride in separate chambers. Both antidotes will be administered through a single needle.

7-3. Classification of Chemical and Biological Agents

a. Chemical agents are classified according to the primary physiological effects they produce, such as blistering, choking, vomiting, and incapacitating agents.

b. Biological warfare agents are classified according to the effect they have on man. The effects include their ability to incapacitate and cause death. Most biological warfare agents are delivered as aerosols that effect the respiratory tract; some can be delivered by releasing infected insects, by contaminating food and water, and by injection (injecting material in individuals by terrorist, not mass exposure). These agents are found in living organisms such as fungi, bacteria, and viruses.

WARNING

Swallowing water or food contaminated with nerve, blister, and other chemical agents and with some biological agents can be fatal. NEVER consume water or food that is suspected of being contaminated until it has been tested and found safe for consumption by medical personnel.

7-4. Conditions for Masking Without Order or Alarm

a. Once an attack with a chemical or biological agent is detected or suspected, or information is available that such an agent is about to be used, you must **STOP BREATHING** and mask immediately. **DO NOT WAIT** to receive an order or alarm under the following circumstances:

- Your position is hit by artillery missiles, rockets that produce vapors, smoke, and mists, and aerial sprays.
- Smoke or vapor cloud from an unknown source is present or approaching.
- A suspicious odor, liquid, or solid is present.
- A chemical or biological warfare agent attack is occurring.
- You are entering an area known or suspected of being contaminated.
- When casualties are being received from an area where chemical or biological agents have reportedly been used.
- You have one or more of the following symptoms:
  - An unexplained runny nose.
  - A sudden unexplained headache.
  - A feeling of choking or tightness in the chest or throat.
  - Dimness of vision.
  - Irritation of the eyes.
  - Difficulty in or increased rate of breathing without obvious reasons.
  - Sudden feeling of depression.
  - Dread, anxiety, or restlessness.
  - Dizziness or light-headedness.

- Slurred speech.
- Unexplained laughter or unusual behavior is noted in others.
- Numerous unexplained ill personnel.
- Service members suddenly collapsing without evident cause.
- Animals or birds exhibiting unusual behavior or suddenly dying.

b. For further information on protection and masking procedures, refer to FM 3-4, FM 4-02.7, FM 8-284, and FM 8-285.

7-5. First Aid for a Chemical Attack

Your field protective mask gives protection against biological and chemical warfare agents as well as radiological fallout. With practice you can mask in 9 seconds or less, or put on your mask with hood within 15 seconds.

a. Stop breathing. Don your mask, seal it properly, and clear and check it; then resume breathing. Give the alarm, and continue the mission. Keep your mask on until the "all clear" signal has been given.

NOTE

Keep your mask on until the area is no longer hazardous and you are told to unmask.

b. If symptoms of nerve agent poisoning (paragraph 7-7) appear, immediately give yourself one MARK I or ATNAA.

CAUTION

Do not inject a nerve agent antidote until you are sure you need it.

c. If your eyes and face become contaminated, you must immediately try to get under cover. You need shelter to prevent further contamination while performing decontamination procedures on your face. If no overhead cover is available, put your poncho over your head before beginning the decontamination process. Then you put on the remaining protective clothing. If vomiting occurs, the mask should be lifted momentarily and drained—with your eyes closed and while holding your breath—then replaced, cleared, and sealed.

d. If nerve agents are used, mission permitting, watch for persons needing nerve agent antidotes and immediately follow procedures outlined in paragraph 7-8*b* or *c*.

e. Decontaminate your skin immediately and clothing and equipment as soon as the mission permits.

7-6. Background Information on Nerve Agents

a. Nerve agents are among the deadliest of chemical agents. Nerve agents enter the body by inhalation, by ingestion, and through the skin. Depending on the route of entry and the amount, nerve agents can produce injury or death within minutes. Nerve agents can achieve their effects with small amounts. Nerve agents are absorbed rapidly, and the effects are felt immediately upon entry into the body. You will be issued three MARK Is or three ATNAAs and one Convulsant Antidote for Nerve Agent (CANA). Each MARK I consists of one atropine autoinjector and one pralidoxime chloride (2 PAM Cl) autoinjector (Figure 7-2A). Each ATNAA consist of a multichambered autoinjector with the atropine and pralidoxime chloride in separate chambers (Figure 7-2C). The CANA is a single autoinjector with flanges (Figure 7-2B). Procedures for use of both the MARK I and ATNAA are described below. You will use either the MARK I or the ATNAA in self-aid and buddy aid as issued.

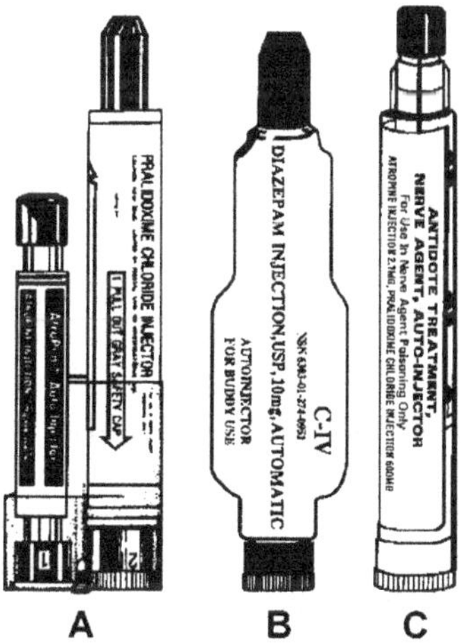

Figure 7-2. Nerve Agent Antidote Kit, MARK I, CANA, and ATNAA.

b. When you have the signs and symptoms of nerve agent poisoning, you should immediately put on the protective mask and then inject yourself with one set of the MARK I or ATNAA. Do not administer the CANA. You should inject yourself in the outer (lateral) thigh muscle (Figure 7-3) or if you are thin, in the upper outer (lateral) part of the buttocks (Figure 7-4).

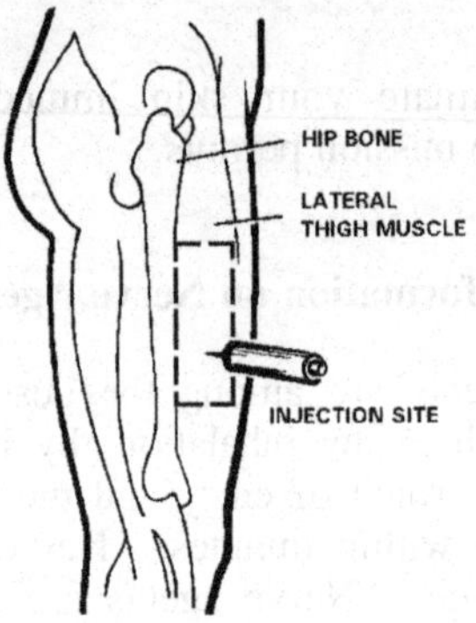

Figure 7-3. Thigh injection site.

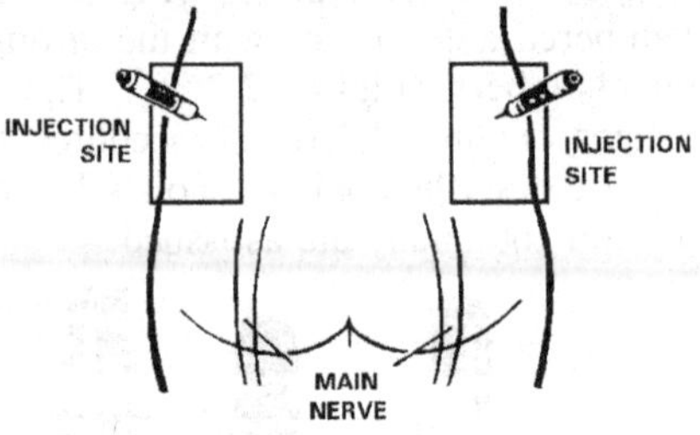

Figure 7-4. Buttocks injection site.

c. Also, you may come upon an unconscious chemical agent casualty who will be unable to care for himself and who will require first aid. You should be able to successfully—

(1) Mask him if he is unmasked.

(2) Inject him, if necessary, with all of **HIS** autoinjectors.

(3) Decontaminate his skin.

(4) Seek medical assistance.

7-7. Signs and Symptoms of Nerve Agent Poisoning

The symptoms of nerve agent poisoning are grouped as **MILD**—those that you recognize and for which you can perform self-aid, and **SEVERE**—those which require buddy aid.

a. MILD Signs and Symptoms.

- Unexplained runny nose.
- Unexplained sudden headache.
- Sudden drooling.
- Difficulty seeing (dimness of vision and miosis).
- Tightness in the chest or difficulty in breathing.
- Localized sweating and muscular twitching in the area of contaminated skin.
- Stomach cramps.
- Nausea.
- Tachycardia followed by bradycardia. (*Tachycardia* is an abnormally rapid heartbeat with a heart rate of over 100 beats per minute. *Bradycardia* is a slow heart rate of less than 60 beats per minute.)

b. SEVERE Signs and Symptoms.

- Strange or confused behavior.
- Wheezing, dyspnea (difficulty in breathing), and coughing.
- Severely pinpointed pupils.
- Red eyes with tearing.
- Vomiting.
- Severe muscular twitching and general weakness.
- Involuntary urination and defecation.

- Convulsions.
- Unconsciousness.
- Respiratory failure.
- Bradycardia.

7-8. First Aid for Nerve Agent Poisoning

First aid for nerve agent poisoning consists of administering the MARK I or ATNAA and CANA.

a. Injection Site. The injection site for administering the antidotes is normally in the outer thigh muscle. The thigh injection site is the area about a hand's width above the knee to a hand's width below the hip joint (Figure 7-3). It is important that the injection be given into a large muscle area. If the individual is thinly built, then the injections should be administered into the upper outer quarter (quadrant) of the buttock (Figure 7-4). Injecting in the buttocks of a thinly built individual avoids injury to the thighbone.

b. Self-Administer MARK I. If you experience any or all of the nerve agent **MILD** symptoms (paragraph 7-7*a*), you must **IMMEDIATELY** put on your protective mask and self-administer one MARK I (Figure 7-2A). Follow the procedure given in Table 7-1. The MARK I is carried in your protective mask carrier, pocket of the MOPP overgarment, or other location as specified in your unit tactical standing operating procedure (TSOP). (In cold weather, the MARK I should be stored in an inside pocket of your clothing to protect the antidote from freezing. A frozen MARK I cannot be immediately used to provide you with antidote, when needed. (However, the MARK I can still be used after complete thawing.)

Table 7-1. Self Aid for Nerve Agent Poisoning

MARK I*	ATNAA*
STEP 1. OBTAIN ONE MARK I.**	STEP 1. OBTAIN ONE ATNAA.**
STEP 2. CHECK INJECTION SITE.	STEP 2. CHECK INJECTION SITE.
STEP 3. HOLD MARK I AT EYE LEVEL WITH NONDOMINANT HAND WITH THE LARGE INJECTOR ON TOP (FIGURE 7-5A).	STEP 3. HOLD ATNAA WITH DOMINANT HAND (FIGURE 7-12A).

Table 7-1. Self Aid for Nerve Agent Poisoning (Continued)

MARK I*	ATNAA*
STEP 4. GRASP SMALL INJECTOR (ATROPINE) (FIGURE 7-5B) AND REMOVE FROM CLIP (FIGURE 7-5C).	STEP 4. GRASP SAFETY CAP WITH NONDOMINANT HAND AND REMOVE FROM INJECTOR (FIGURE 7-12B).
STEP 5. CLEAR HARD OBJECTS FROM INJECTION SITE.	STEP 5. CLEAR HARD OBJECTS FROM INJECTION SITE.
STEP 6. INJECT ATROPINE AT INJECTION SITE APPLYING EVEN PRESSURE TO THE INJECTOR (FIGURE 7-6 OR 7-7). HOLD IN PLACE FOR 10 SECONDS.	STEP 6. INJECT ATNAA AT INJECTION SITE APPLYING EVEN PRESSURE TO THE INJECTOR (FIGURE 7-14 OR 7-15). HOLD IN PLACE FOR 10 SECONDS.
STEP 7. HOLD USED INJECTOR WITH NONDOMINANT HAND.	STEP 7. BEND NEEDLE OF USED INJECTOR BY PRESSING ON A HARD SURFACE TO FORM A HOOK.
STEP 8. GRASP THE LARGE (2 PAM CI) INJECTOR (FIGURE 7-8B) AND PULL IT FROM CLIP (FIGURE 7-8C). DROP CLIP TO GROUND.	STEP 8. ATTACH USED INJECTOR TO BLOUSE POCKET FLAP OF BDO/JSLIST (FIGURE 7-16).
STEP 9. INJECT 2 PAM CI AT INJECTION SITE APPLYING EVEN PRESSURE TO THE INJECTOR (FIGURE 7-6 OR 7-7). HOLD IN PLACE FOR 10 SECONDS.	STEP 9. MASSAGE INJECTION SITE, MISSION PERMITTING.
STEP 10. BEND THE NEEDLES OF ALL USED INJECTORS BY PRESSING ON A HARD SURFACE TO FORM A HOOK.	
STEP 11. ATTACH ALL USED INJECTORS TO BLOUSE POCKET FLAP OF BDO/JSLIST (FIGURE 7-9).	
STEP 12. MASSAGE INJECTION SITE, MISSION PERMITTING.	

* USE STEPS LISTED FOR TYPE OF ANTIDOTE DEVICE ISSUED.
** ONLY ADMINISTER ONE MARK I OR ATNAA AS SELF-AID. DO NOT SELF-ADMINISTER CANA.

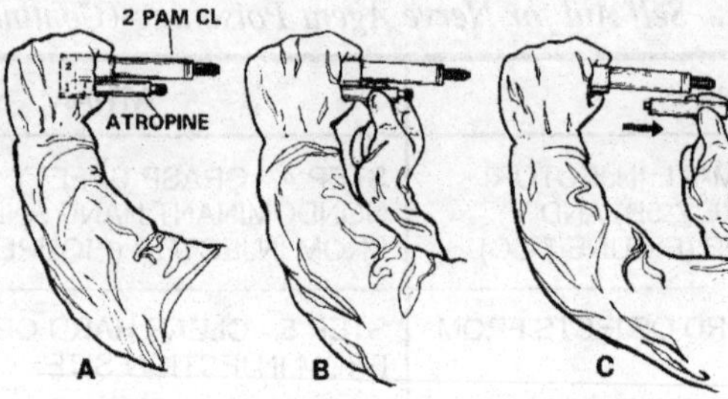

Figure 7-5. Removing the atropine autoinjector from the MARK I clip.

CAUTION

DO NOT cover or hold the needle end with your hand, thumb, or fingers—you might accidentally inject yourself. An accidental injection into the hand **WILL NOT** deliver an effective dose of the antidote, especially if the needle goes through the hand.

Figure 7-6. Thigh injection site for self-aid.

NOTE

If you are thinly built, inject yourself into the upper outer quadrant of the buttock (Figure 7-7). There is a nerve that crosses the buttocks; hitting this nerve can cause paralysis. Therefore, you must only inject into the *upper outer quadrant* of the buttock.

Figure 7-7. Buttocks injection site for self-aid.

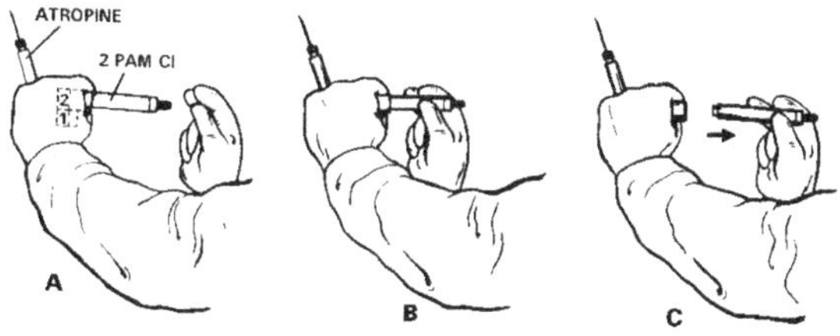

Figure 7-8. Removing the 2 PAM Cl autoinjector from the MARK I clip.

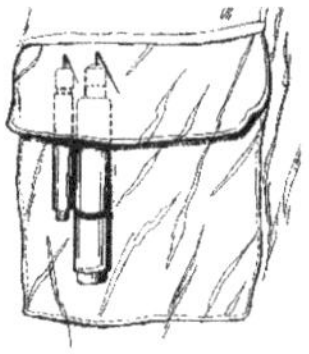

Figure 7-9. One set of used MARK I autoinjectors attached to pocket flap.

NOTES

1. **DO NOT** give yourself another set of injections. If you are able to walk without assistance, know who you are, and where you are, you **WILL NOT** need the second set of injections. (If not needed, giving yourself a second set of MARK I injections or ATNAA may create a nerve agent antidote overdose, which could cause incapacitation [inability to perform mission or defend yourself].)

2. If you continue to have symptoms of nerve agent poisoning, seek someone else (a buddy) to check your symptoms and administer the additional sets of injections, if required.

c. Buddy Evaluation and Buddy Aid. Service members may seek assistance after self-aid (self-administering one MARK I or ATNAA) or may become incapacitated after self-aid. A buddy must evaluate the individual to determine if additional antidotes are required to counter the effects of the nerve agent. Also, service members may experience **SEVERE** symptoms of nerve agent poisoning (paragraph 7-7*b*); they will not be able to treat themselves. In either case, other service members must perform buddy aid as quickly as possible. Before initiating buddy aid, determine if one set of MARK I autoinjectors has already been used so that no more than three sets of the antidote are administered. Buddy aid also includes administering the CANA with the third MARK I or ATNAA to prevent convulsions. Follow the procedures indicated in Table 7-2.

WARNING

Squat, DO NOT kneel, when masking the casualty or administering the nerve agent antidote to the casualty. Kneeling may force the chemical agent into or through your protective clothing.

CAUTION

DO NOT use your own MARK I, ATNAA, or CANA on a casualty. If you use your own, you may not have any antidote if needed for self-aid.

WARNING

DO NOT inject into areas close to the hip, knee, or thigh-bone.

Table 7-2. Buddy Aid/Combat Lifesaver Aid for Nerve Agent Casualty.

MARK I*	ATNAA*	CANA**
STEP 1. MASK THE CASUALTY AND POSITION HIM ON HIS SIDE (SWIMMER'S POSITION).	STEP 1. MASK THE CASUALTY AND POSITION HIM ON HIS SIDE (SWIMMER'S POSITION).	STEP 1. OBTAIN BUDDY'S CANA.
STEP 2. POSITION YOURSELF NEAR THE CASUALTY'S THIGH.	STEP 2. POSITION YOURSELF NEAR THE CASUALTY'S THIGH.	STEP 2. CHECK INJECTION SITE.

Table 7-2. Buddy Aid/Combat Lifesaver Aid for Nerve Agent Casualty (Continued).

MARK I*	ATNAA*	CANA**
STEP 3. OBTAIN BUDDY'S THREE OR REMAINING MARK Is.	STEP 3. OBTAIN BUDDY'S THREE OR REMAINING ATNAAs.	STEP 3. HOLD CANA IN A CLOSED FIST WITH DOMINANT HAND (FIGURE 7-12A).
STEP 4. CHECK INJECTION SITE.	STEP 4. CHECK INJECTION SITE.	STEP 4. GRASP SAFETY CAP WITH NONDOMINANT HAND AND REMOVE FROM INJECTOR (FIGURE 7-12B).
STEP 5. HOLD MARK I WITH NONDOMINANT HAND (FIGURE 7-5A).	STEP 5. HOLD ATNAA IN A CLOSED FIST WITH DOMINANT HAND (FIGURE 7-12A).	STEP 5. CLEAR HARD OBJECTS FROM INJECTION SITE.
STEP 6. GRASP SMALL INJECTOR (ATROPINE) AND REMOVE FROM CLIP (FIGURE 7-5B).	STEP 6. GRASP SAFETY CAP WITH NONDOMINANT HAND AND REMOVE FROM INJECTOR (FIGURE 7-12B).	STEP 6. INJECT CANA AT INJECTION SITE BY APPLYING EVEN PRESSURE TO THE INJECTOR, NOT A JABBING MOTION (FIGURE 7-14 OR 7-15). HOLD IN PLACE FOR 10 SECONDS.
STEP 7. CLEAR HARD OBJECTS FROM INJECTION SITE.	STEP 7. CLEAR HARD OBJECTS FROM INJECTION SITE.	STEP 7. BEND NEEDLE OF INJECTOR BY PRESSING ON A HARD SURFACE TO FORM A HOOK.
STEP 8. INJECT ATROPINE AT INJECTION SITE BY APPLYING EVEN PRESSURE TO THE INJECTOR, NOT A JABBING MOTION (FIGURE 7-10 OR 7-11). HOLD IN PLACE FOR 10 SECONDS.	STEP 8. INJECT ATNAA AT INJECTION SITE BY APPLYING EVEN PRESSURE TO THE INJECTOR, NOT A JABBING MOTION (FIGURE 7-14 OR 7-15). HOLD IN PLACE FOR 10 SECONDS.	STEP 8. ATTACH USED INJECTOR TO BLOUSE POCKET FLAP OF BDO/JSLIST (FIGURE 7-16).
STEP 9. HOLD USED INJECTOR BETWEEN LITTLE FINGER AND RING FINGER OF NONDOMINANT HAND (FIGURE 7-5A).	STEP 9. BEND NEEDLE OF INJECTOR BY PRESSING ON A HARD SURFACE TO FORM A HOOK.	STEP 9. MASSAGE INJECTION SITE, MISSION PERMITTING.
STEP 10. PULL LARGE INJECTOR (2 PAM CI) FROM CLIP (FIGURE 7-5C). DROP CLIP TO GROUND.	STEP 10. ATTACH ALL USED INJECTORS TO BLOUSE POCKET FLAP OF BDO/JSLIST (FIGURE 7-16).	

Table 7-2. Buddy Aid/Combat Lifesaver Aid for Nerve Agent Casualty (Continued).

MARK I*	ATNAA*	CANA**
STEP 11. INJECT 2 PAM CI AT INJECTION SITE BY APPLYING EVEN PRESSURE TO THE INJECTOR, NOT A JABBING MOTION (FIGURE 7-10 OR 7-11). HOLD IN PLACE FOR 10 SECONDS.	STEP 11. MASSAGE INJECTION SITE, MISSION PERMITTING.	
STEP 12. REPEAT STEPS ABOVE FOR REMAINING MARK Is.		
STEP 13. BEND THE NEEDLES OF ALL USED INJECTORS BY PRESSING ON A HARD SURFACE TO FORM A HOOK.		
STEP 14. ATTACH ALL USED INJECTORS TO BLOUSE POCKET FLAP OF BDO/JSLIST (FIGURE 7-13).		
STEP 15. MASSAGE INJECTION SITE, MISSION PERMITTING.		

* USE STEPS LISTED FOR TYPE OF ANTIDOTE DEVICE ISSUED.

** CANA IS USED IN BUDDY AID/CLS AID ONLY. DO NOT USE IN SELF-AID.

NOTE

If the casualty is thinly built, inject the antidote into the buttock. Only inject the antidote into the upper outer portion of the casualty's buttock (Figure 7-11). This avoids hitting the nerve that crosses the buttocks (Figure 7-4). Hitting this nerve can cause paralysis.

Figure 7-10. Injecting the casualty's thigh (*Mark I or CANA*).

Figure 7-11. Injecting the casualty's buttocks (*Mark I or CANA*).

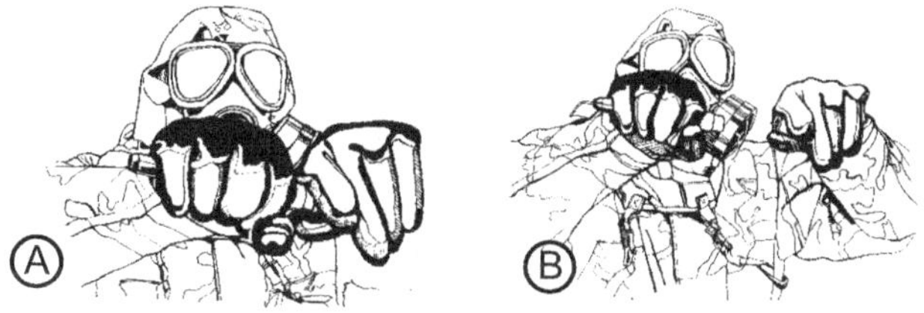

Figure 7-12. Preparing CANA or ATNAA for injection.

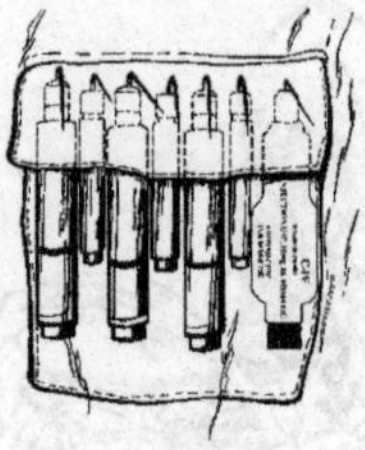

Figure 7-13. Three sets of used MARK I autoinjectors and one CANA autoinjector attached to pocket flap.

d. Self-Administer Antidote Treatment Nerve Agent Autoinjector. If you experience any or all of the nerve agent **MILD** symptoms (paragraph 7-7*b*), you must **IMMEDIATELY** self-administer one ATNAA following the procedure given Table 7-1.

NOTE

If you are thinly-built, inject yourself into the upper outer quarter (quadrant) of the buttock (Figure 7-15). There is a nerve that crosses the buttocks; hitting this nerve can cause paralysis. Therefore, you must only inject into the upper outer quarter (quadrant) of the buttocks.

Figure 7-14. Self-administration of ATNAA (thigh).

Figure 7-15. Self-administration of ATNAA (buttock).

NOTE

If you continue to have symptoms of nerve agent poisoning, seek someone else (a buddy) to check your symptoms and administer your remaining sets of injections, if required.

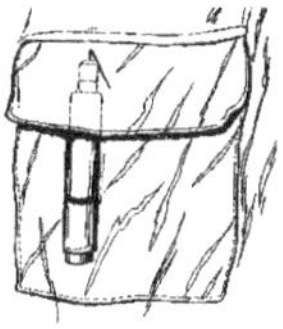

Figure 7-16. Used ATNAA attached to clothing.

e. Buddy Assistance. Service members may seek assistance after self-aid (self-administering one ATNAA) or may become incapacitated after self-aid. A buddy must evaluate the individual to determine if additional antidotes are required to counter the effects of the nerve agent. Also, service members may experience **SEVERE** symptoms of nerve agent poisoning (paragraph 7-7*b*); they will not be able to treat themselves. In either case, other service members must perform buddy aid as quickly as possible. Before initiating buddy aid, determine if one ATNAA has already been used so that no more than three ATNAA are administered. Buddy aid also includes administering the CANA with the third ATNAA to prevent convulsions. Follow the procedures indicated in Table 7-2.

WARNING

Squat, DO NOT kneel, when masking the casualty or administering the nerve agent antidotes to the casualty. Kneeling may force any chemical agent on your overgarment into or through your protective clothing.

Figure 7-17. Buddy injecting casualty's outer thigh (ATNAA or CANA).

NOTE

If the casualty is thinly built, inject the antidote into the buttocks (Figure 7-18). Only inject the antidote into the upper outer portion of the casualty's buttocks. This avoids hitting the nerve that crosses the buttocks (Figure 7-4). Hitting this nerve can cause paralysis.

WARNING

DO NOT inject into areas close to the hip, knee, or thighbone.

Figure 7-18. Buddy injecting casualty's buttocks (ATNAA or CANA).

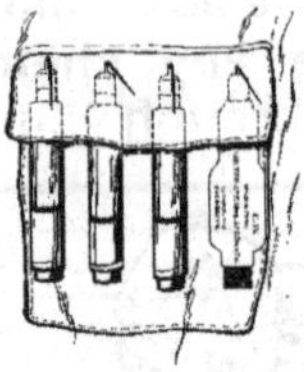

Figure 7-19. Three used ATNAAs and one CANA autoinjector attached to clothing.

f. *Combat Lifesaver.*

(1) The combat lifesaver must check to verify if the individual has received three sets of MARK I or ATNAAs. If not, the combat lifesaver performs first aid as described for buddy aid above. If the

individual has received the initial three sets of MARK I, then the combat lifesaver may administer additional atropine injections at approximately 15 minute intervals until atropinization is achieved (that is a heart rate above 90 beats per minute, reduced bronchial secretions, and reduced salivations). Administer additional atropine at intervals of 30 minutes to 4 hours to maintain atropinization or until the casualty is placed under the care of medical personnel. Check the heart rate by lifting the casualty's mask hood and feeling for a pulse at the carotid artery. Request medical assistance as soon as the tactical situation permits.

(2) The CLS should administer additional CANA to casualties suffering convulsions. Administer a second, and if needed, a third CANA at 5 to 10 minute intervals for a maximum of three injections (30 milligrams diazepam). Follow the steps and procedures described in buddy aid for administering the CANA. **DO NOT** give more than two additional injections for a total of three (one self-aid plus two by the CLS).

7-9. Blister Agents

Blister agents (vesicants) include mustard (H and HD), nitrogen mustards (HN), lewisite (L), and other arsenicals, mixtures of mustards and arsenicals, and phosgene oxime (CX). Blister agents may act on the eyes, mucous membranes, lungs, and skin. They burn and blister the skin or any other body parts they contact. Even relatively low doses may cause serious injury. Blister agents damage the respiratory tract (nose, sinuses, and windpipe) when inhaled and cause vomiting and diarrhea when absorbed. Lewisite and CX cause immediate pain on contact. However, mustard agents are deceptive as there is little or no pain at the time of exposure. Thus, in some cases, signs of injury may not appear for several hours after exposure.

a. Protective Measures. Your protective mask with hood and protective overgarment provide protection against blister agents. If it is known or suspected that blister agents are being used, **STOP BREATHING**, put on your mask and your protective overgarment.

CAUTION

Large drops of liquid vesicants on the protective overgarment ensemble may penetrate it if allowed to stand for an extended period. Remove large drops as soon as possible.

b. *Signs and Symptoms of Blister Agent Poisoning.*

(1) Immediate and intense pain upon contact with L, LH (lewisite and mustard) mixture, and CX. No initial pain upon contact with mustard.

(2) Inflammation and blisters (burns) resulting in tissue destruction. The severity of a chemical burn is directly related to the concentration of the agent and the duration of contact with the skin. The longer the agent is in contact with the tissue, the more serious the injury will be.

(3) Vomiting and diarrhea. Exposure to high concentrations of vesicants may cause vomiting or diarrhea.

(4) Death. The blister agent vapors absorbed during ordinary field exposure will probably not cause enough internal body (systemic) damage to result in death. However, death may occur from prolonged exposure to high concentrations of vapor or from extensive liquid contamination over wide areas of the skin, particularly *when decontamination is neglected or delayed.*

c. *First Aid Measures.*

(1) Use your M291 Skin Decontaminating Kit to decontaminate your skin and use water to flush contaminated eyes. Decontamination of vesicants must be done immediately (within 1 minute is best).

(2) If blisters form, cover them loosely with a field dressing and secure the dressing.

CAUTION

Blisters are actually burns. **DO NOT** attempt to decontaminate the skin where blisters have formed, as the agent has already been absorbed.

(3) If you receive blisters over a wide area of the body, you are considered seriously burned. Seek medical assistance immediately.

(4) If vomiting occurs, the mask should be lifted momentarily and drained—while the eyes are closed and the breath is held—and replaced, cleared, and sealed.

(5) Remember, if vomiting or diarrhea occurs after having been exposed to blister agents, seek medical assistance immediately.

7-10. Choking Agents (Lung-Damaging Agents)

Chemical agents that attack lung tissue, primarily causing fluid buildup (pulmonary edema), are classified as choking agents (lung-damaging agents). This group includes phosgene (CG), diphosgene (DP), chlorine (Cl), and chloropicrin (PS). Of these four agents, CG is the most dangerous and is more likely to be employed by the enemy in future conflict.

a. *Protective Measures.* Your protective mask gives adequate protection against choking agents.

b. *Signs and Symptoms.* During and immediately after exposure to choking agents (depending on agent concentration and length of exposure), you may experience some or all of the following signs and symptoms:

- Tears (lacrimation).
- Coughing.
- Choking.
- Tightness of chest.
- Nausea and vomiting.
- Headaches.

c. *Self-Aid.*

(1) The protective mask should be put on immediately when any of the conditions described in *b* above exist. Another indication of a CG attack is an odor like newly mown hay; however, **DO NOT** rely upon odor as indication of a chemical attack.

(2) If some CG is inhaled, normal combat duties should be continued unless there is difficulty in breathing, nausea, vomiting, or more than the usual shortness of breath during exertion. If any of the above symptoms occur and the mission permits, remain at quiet rest until medical evacuation is accomplished.

d. *Death.* With ordinary field exposure to choking agents, death will probably not occur. *However, prolonged exposure to high concentrations of the vapor and neglect or delay in masking can be fatal.*

7-11. Cyanogen (Blood) Agents

Cyanogen agents interfere with proper oxygen utilization in the body. Hydrogen cyanide (AC) and cyanogen chloride (CK) are the primary agents in this group.

a. *Protective Measures*. Your protective mask with a fresh filter gives adequate protection against field concentrations of cyanogen agent vapor. The protective overgarments, as well as the mask, are needed when exposed to liquid AC.

b. *Signs and Symptoms*. During and immediately after exposure to cyanogen agents (depending on agent concentration and length of exposure), you may experience some or all of the following signs and symptoms:

- Tearing (lacrimation).
- Eye, nose, and throat irritation.
- Sudden stimulation of breathing (unable to hold breath).
- Nausea.
- Coughing.
- Tightness of chest.
- Headache.
- Light-headedness (dizziness).
- Unconsciousness.

c. *First Aid*.

(1) *Hydrogen cyanide*. During any chemical attack, if you get a sudden stimulation of breath or detect an odor like bitter almonds, **PUT ON YOUR MASK IMMEDIATELY**. Speed is absolutely essential since this agent acts so rapidly that within a few seconds its effects will make it impossible for service members to put on their mask by themselves. Stop breathing until the mask is on, if at all possible. This may be very difficult since the agent strongly stimulates respiration.

(2) *Cyanogen chloride*. Put your mask on immediately if you experience any irritation of the eyes, nose, or throat. Service members

who are unable to mask should be masked by the nearest service member (buddy).

d. *Medical Assistance.* If you suspect that you have been exposed to blood agents, seek medical assistance immediately.

7-12. Incapacitating Agents

An incapacitating agent is a chemical agent which produces temporary, disabling conditions which persist for hours to days after exposure. Unlike riot control agents, which usually are momentary or fleeting in action, incapacitating agents have a persistent effect. It is likely that smoke-producing munitions or aerosols will disseminate such agents, thus making breathing their means of entry into the body. The protective mask is, therefore, essential.

a. There are no specific first aid measures to relieve the symptoms of incapacitating agents. Supportive first aid and physical restraint may be indicated. If the casualty is stuporous or comatose, be sure that respiration is unobstructed; then turn him on his side in case vomiting should occur. Complete cleansing of the skin with soap and water should be done as soon as possible; or, the M291 Skin Decontaminating Kit can be used if washing is impossible. Remove weapons and other potentially harmful items from service members who are suspected of having these symptoms. Harmful items include cigarettes, matches, medications, and small items that might be swallowed accidentally. Delirious (confused) persons have been known to attempt to eat items bearing only a superficial resemblance to food.

b. Incapacitating agents (anticholinergic drugs BZ type) may produce alarming dryness and coating of the lips and tongue; however, there is usually no danger of immediate dehydration. Fluids should be given sparingly, if at all, because of the danger of vomiting and because of the likelihood of temporary urinary retention due to paralysis of bladder muscles.

c. If the body temperature is elevated and mucous membranes are dry, immediate and vigorous cooling (as for heatstroke) is indicated. Methods that can be used to cool the skin are spraying with cool water and air circulation (fanning); applying alcohol soaked cloths and air circulation; and providing maximum exposure to air in a shaded area, along with maximum air circulation. Such cases are usually a result of anticholinergic poisoning. Rapid evacuation should be accomplished since medical treatment with the appropriate medication may be lifesaving.

CAUTION

DO NOT use **ice** for cooling the skin.

d. Reassurance and a firm, but friendly, attitude by individuals providing first aid will be beneficial if the casualty appears to comprehend what is being said. Conversation is a waste of time if the service member is incoherent or cannot understand what is being said. In such cases, the less said, the better it is—these casualties will benefit more from prompt and vigorous restraint and evacuation to an MTF.

7-13. Incendiaries

Incendiaries can be grouped as WP, thickened gasoline, metal, and oil and metal. You must learn to protect yourself against these incendiaries.

a. White phosphorus is used primarily as a smoke producer but can be used for its incendiary effect to ignite field expedients and combustible materials. The burns from WP are usually multiple, deep, and variable in size. When particles of WP get on the skin or clothing, they continue to burn until deprived of air. They also have a tendency to stick to a surface and must be brushed off or picked out.

(1) If burning particles of WP strike and stick to your clothing, quickly take off the contaminated clothing before the WP burns through to the skin.

(2) If burning WP strikes your skin, smother the flame with water, a wet cloth, or mud.

NOTE

Since WP is soluble in oil, **DO NOT** use grease, oily ointments, or eye ointments to smother the flame.

(3) Keep the WP particles covered with a wet material to exclude air until you can remove them or have them removed from your skin.

(4) Remove the WP particles from the skin by brushing them with a wet cloth and by picking them out with a knife, bayonet, stick, or other available object.

(5) Seek medical assistance when the mission permits.

b. Thickened fuel mixtures (napalm) have a tendency to cling to clothing and body surfaces, thereby producing prolonged exposure and severe burns. The first aid for these burns is the same as for other heat burns. The heat and irritating gases given off by these combustible mixtures may cause lung damage, which must be treated by medical personnel.

c. Metal incendiaries pose special problems. Thermite particles on the skin should be immediately cooled with water and then removed. The first aid for these burns is the same as for other heat burns. Particles of magnesium on the skin burn quickly and deeply. Like other metal incendiaries, they must be removed. Ordinarily, medical personnel should do the complete removal of these particles as soon as possible. Immediate medical treatment is required.

d. Oil and metal incendiaries have much the same effect on contact with the skin and clothing as those discussed (*b* and *c* above). First aid measures for burns are discussed in Chapter 3.

7-14. Biological Agents and First Aid

a. Biological attacks can result in combat ineffectiveness by introducing disease-causing organisms into a troop population.

b. Once a disease is identified, first aid or medical treatment is initiated, depending on the seriousness of the disease. First aid measures are concerned with observable symptoms of the disease such as diarrhea or vomiting.

7-15. Toxins

Toxins are alleged to have been used in past conflicts. Witnesses and victims have described the agent as toxic rain (or yellow rain) because it was reported to have been released from aircraft as a yellow powder or liquid that covered ground, structures, vegetation, and people.

a. Signs and Symptoms. The occurrence of the symptoms from toxins may appear in a period of a few minutes to several hours depending on the particular toxin, the service member's susceptibility, and the amount of toxin inhaled, ingested, or deposited on the skin. Symptoms from toxins usually involve the central nervous system but are often preceded by less prominent symptoms, such as nausea, vomiting, diarrhea, cramps, or stomach irritation and burning sensation. Typical neurological symptoms often develop rapidly in severe cases; for example, visual disturbances, inability to swallow, speech difficulty, lack of muscle coordination, and sensory abnormalities (numbness of mouth, throat, or extremities). Yellow rain (mycotoxins) also may have hemorrhagic symptoms, which could include any or all of the following:

- Dizziness.
- Severe itching or tingling of the skin.
- Formation of multiple, small, hard blisters.
- Coughing up blood.
- Shock (which could result in death).

b. Self-Aid. Upon recognition of an attack employing toxins, you must immediately take the following actions:

(1) Stop breathing, put on your protective mask with hood, and then resume breathing. Next, put on your protective clothing.

(2) Should severe itching of the face become unbearable, quickly—

- Loosen the cap on your canteen.
- Take and hold a deep breath and lift your mask.
- While holding your breath, close your eyes and flush your face with generous amounts of water.

CAUTION

DO NOT rub or scratch your eyes. Try not to let the water run onto your clothing or protective overgarment.

- Put your protective mask back on, seat it properly, clear it, and check it for a seal; then resume breathing.
- Decontaminate your skin by bathing with soap and water as soon as the mission permits.

- Change clothing and decontaminate your protective mask using soap and water. Replace the filters if directed.

(3) If vomiting occurs, the mask should be lifted momentarily and drained—while the eyes are closed and the breath is held—and replaced, cleared, and sealed.

c. *Medical Assistance.* If you suspect that you have been exposed to toxins, you should seek medical assistance immediately.

7-16. Nuclear Detonation

a. Three types of injuries may result from a nuclear detonation. These are thermal, blast, and radiation injuries. Many times the casualty will have a combination of these types of injuries. First aid for thermal and blast injuries is provided based on observable injuries, such as burns, hemorrhage, or fractures.

b. The signs and symptoms of radiation illness in the initial phase include the rapid onset of nausea, vomiting, and malaise (tiredness). The only first aid procedure for radiological casualties is decontamination.

CHAPTER 8

FIRST AID FOR PSYCHOLOGICAL REACTIONS

8-1. General

Psychological first aid is as natural and reasonable as physical first aid and is just as familiar. When you were hurt as a child, the understanding attitude of your parents did as much as the psychological effect of a bandage. Later, your disappointment or grief was eased by supportive words from a friend. Certainly, taking a walk and talking things out with a friend are familiar ways of dealing with an emotional crisis. The same natural feelings that make us want to help a person who is injured make us want to give a helping hand to a buddy who is upset. *Psychological first aid* really means nothing more complicated than assisting people with emotional distress whether it results from physical injury, disease, or excessive stress. Emotional distress is not always as visible as a wound or a broken bone. However, overexcitement, severe fear, excessive worry, deep depression, misdirected irritability, and anger are signs that stress has reached the point of interfering with effective coping. The more noticeable the symptoms become, the more urgent the need for you to be of help and the more important it is for you to know *how* to help.

8-2. Importance of Psychological First Aid

You must know how to give psychological first aid to be able to help yourself, your buddies, and your unit in order to keep performing the mission. Psychological first aid measures are simple and easy to understand. Your decision of what to do depends upon your ability to observe the service member and understand his needs. Making the best use of resources requires ingenuity on your part. A stress reaction resulting in poor judgment can cause injury or even death to yourself or others on the battlefield. It can be even more dangerous if other persons are affected by the judgment of an emotionally upset service member. If it is detected early enough, the affected service member stands a good chance of remaining in his unit as an effective member. If it is not detected early and if the service member becomes more emotionally upset, he may become a threat to himself and to others.

8-3. Situations Requiring Psychological First Aid

- Psychological first aid (buddy aid) is most needed at the first sign that a service member cannot perform the mission because of emotional distress. Stress is inevitable in combat, in hostage and terrorist situations, and in civilian disasters such as floods, hurricanes, or industrial accidents. Most emotional reactions to such situations are temporary, and the service member can still carry on with encouragement. Painful or disruptive symptoms may last for minutes, hours, or days. However, if the stress symptoms are seriously disabling, they may be psychologically contagious and endanger not only the emotionally upset service member but also the entire unit.

- Sometimes people continue to function well during a disastrous event, but suffer from emotional scars which impair their job performance or quality of life at a later time. Painful memories and dreams may recur for months and years and still be considered a normal reaction. However, if the memories are so painful that the person must avoid all situations which arouse them, becomes socially withdrawn, or shows symptoms of anxiety, depression, or substance abuse, he needs treatment. Experience with police, firemen, emergency medical technicians, and others who deal with disasters has proved that the routine application of psychological first aid to all the participants, including those who have functioned well, greatly reduces the likelihood of future serious post-traumatic stress disorders (PTSDs).

8-4. Interrelationship of Psychological and Physical First Aid

Psychological first aid should go hand in hand with physical first aid. The discovery of a physical injury or cause for an inability to function does not rule out the possibility of a psychological injury (or vice versa). The person suffering from pain, shock, fear of serious injury, or fear of death does not respond well to joking, indifference, or fearful-tearful attention. Fear and anxiety may take as high a toll of the service member's strength as does the loss of blood.

8-5. Goals of Psychological First Aid

The goals of psychological first aid are to—

- Be supportive; assist the service member in dealing with his stress reaction.
- Prevent, and if necessary control, behavior harmful to himself and to others.
- Return the service member to duty as soon as possible after dealing with the stress reaction.

8-6. Respect for Others' Feelings

a. Accept the service member you are trying to help without censorship or ridicule. Respect his right to his own feelings. Even though your feelings, beliefs, and behavior are different, DO NOT blame or make light of him for the way he feels or acts. Your purpose is to help him in this tough situation, not to be his critic. A person DOES NOT WANT to be upset and worried. When he seeks help, he needs and expects consideration of his fears, not abrupt dismissal or ridicule.

b. Realize that people are the products of a wide variety of factors. All people DO NOT react the same way to the same situations. Each individual has complex needs and motivations, both conscious and unconscious, that are uniquely his own. Often the one thing that finally causes the person to become overloaded by a stressful situation is not the stressor itself, but some other problem.

8-7. Emotional and Physical Disability

a. Accept emotional disability as being just as real as physical disability. If a service member's ankle is seriously sprained in a fall, no one expects him to run right away. A service member's emotions may be temporarily strained by the overwhelming stress of battle or other traumatic incident. DO NOT demand that he pull himself together immediately and carry on without a break. Some individuals can pull themselves together immediately, but others cannot. The service member whose emotional stability has been disrupted has a disability just as real as the service member who has sprained his ankle. There is an unfortunate tendency in many people to regard as real only what they can see, such as a wound or bleeding. Some people tend to assume that damage involving a person's mind and emotions is just imagined, that he is not really sick or injured, and that he could overcome his trouble by using his will power.

b. The terms *it's all in your head, snap out of it,* and *get control of yourself* are often used by people who believe they are being helpful. Actually, these terms are expressions of hostility because they show lack of understanding. They only emphasize weakness and inadequacy. Such terms are of no use in psychological first aid.

c. Every physically injured person has some emotional reaction to the fact that he is injured.

(1) It is normal for an injured person to feel upset. The more severe the injury, the more insecure and fearful he becomes, especially

if the injury is to a body part which is highly valued. For example, an injury to the eyes or the genitals, even though relatively minor, is likely to be extremely upsetting. An injury to some other part of the body may be especially disturbing to an individual for his own particular reason. For example, an injury of the hand may be a terrifying blow to a surgeon or an injury to the eye of a pilot.

(2) An injured service member always feels less secure, more anxious, and more afraid not only because of what has happened to him but because of what he imagines may happen as a result of his injury. This fear and insecurity may cause him to be irritable, uncooperative, or unreasonable. As you help him, always keep in mind that such behavior has little or nothing to do with you personally. He needs your patience, reassurance, encouragement, and support.

8-8. Combat and Other Operational Stress Reactions

Stress reaction is a temporary emotional disorder or inability to function, experienced by a previously normal service member as a reaction to the overwhelming or cumulative stress of combat. Stress reaction gets better with reassurance, rest, physical replenishment, and activities that restore confidence. All service members are likely to feel stress reaction under conditions of intense and/or prolonged stress. They may even become stress reaction casualties, unable to perform their mission for hours or days. Other combat and operational stress reactions (COSRs) may result in negative behavior, but are not termed *stress reaction,* as they need more intensive treatment. These negative COSRs may result in misconduct stress behaviors such as drug and alcohol abuse, criminal acts, looting, desertion, and self-inflicted wounds. These harmful COSRs can often be prevented by good psychological first aid. Service members who commit misconduct stress behaviors may require disciplinary action rather than medical treatment.

8-9. Reactions to Stress

Most service members react to stressful incidents after the situation has passed. All service members feel some fear. This fear may be greater than they have experienced at any other time, or they may be more aware of their fear. In such a situation, they should not be surprised if they feel shaky or become sweaty, nauseated, or confused. These reactions are normal and are not a cause for concern. However, some reactions, either short- or long-term, will cause problems if left unchecked. See paragraph 8-13 for more information.

a. Emotional Reactions.

(1) The most obvious combat stress reaction (CSR) is inefficient performance. This can be demonstrated by—

- Slow thinking (or reaction time).
- Difficulty recognizing priorities and seeing what needs to be done.
- Difficulty getting started.
- Indecisiveness and having trouble focusing attention.
- Tendency to do familiar tasks and be preoccupied with familiar details. (This can reach the point where the person is very passive, such as just sitting or wandering about not knowing what to do.)

(2) A less common reaction may be uncontrolled emotional outbursts; this can be demonstrated by crying, screaming, or laughing. Some service members will react in the opposite way. They will be very withdrawn and silent and try to isolate themselves from everyone. These service members should be encouraged to remain with their assigned unit. Uncontrolled reactions may appear by themselves or in any combination (the person may be crying uncontrollably one minute and then laughing the next). In this state, the person is restless and cannot keep still. He may run about, apparently without purpose. Inside, he feels a great rage or fear and his physical acts may show this. In his anger he may indiscriminately strike out at others.

b. *Loss of Adaptability.*

(1) In a desperate attempt to get away from the danger, which has overwhelmed him, a service member may panic and become confused. His mental ability may be so impaired he cannot think clearly or even follow simple commands. His judgment may be faulty and he may not be aware of his actions, such as standing up in his fighting position during an attack.

(2) In other cases, overwhelming stress may produce symptoms that are often associated with head injuries. For example, the service member may appear dazed or be found wandering around aimlessly. He may appear confused and disoriented and may seem to have a complete or partial loss of memory. In such cases, especially when no eyewitnesses can provide evidence that the service member has NOT suffered a head injury, it is necessary for him to be rapidly medically evacuated. **DO NOT** allow the

service member to expose himself to further personal danger until the cause of the problem has been determined.

c. Sleep Disturbance and Repetition of Dreams. A person who has been overwhelmed by stress often has difficulty sleeping. The service member may experience nightmares related to the stressors. Remember that nightmares, in themselves, are not considered abnormal when they occur soon after a period of intensive stress. As time passes, the nightmares usually become less frequent and less intense. In extreme cases, a service member, even when awake, may think repeatedly of the incident, feel as though it is happening again, and act out parts of his stress over and over again. For some persons, this repetitious reexperiencing of the stressful event may be necessary for eventual recovery; therefore, it should not be discouraged or viewed as abnormal. For the person reexperiencing the event, such reaction may be disruptive. The service member needs to be encouraged to *ventilate* about the incident. Ventilation is a technique where the service member is given the opportunity to talk extensively, often repetitiously about the experience.

8-10. Severe Stress or Stress Reaction

You do not need specialized training to recognize severe stress or stress reaction that will cause problems for the service member, the unit, or the mission. Reactions that are less severe, however, are more difficult to detect. To determine whether a person needs help, you must observe him to see whether he is doing something meaningful, performing his duties, taking care of himself, behaving in an unusual fashion, or acting out of character.

8-11. Application of Psychological First Aid

The emotionally disturbed service member has built a barrier against fear. He does this for his own protection, although he is probably not aware that he is doing it. If he finds that he does not have to be afraid and that there are normal, understandable things about him, he will feel safer in dropping this barrier. Persistent efforts to make him realize that you want to understand him will be reassuring, especially if you remain calm. Nothing can cause an emotionally disturbed person to become even more fearful than feeling that others are afraid of him. Try to remain calm. Familiar things, such as a cup of coffee, the use of his name, attention to a minor wound, being given a simple job to do, or the sight of familiar people and activities, will add to his ability to overcome his fear. He may not respond well if you get excited, angry, or abrupt.

a. *Ventilation.* After the service member becomes calmer, he is likely to have dreams about the stressful event. He also may think about it when he is awake or even repeat his personal reaction to the event. One benefit of this natural pattern is that it helps him master the stress by going over it just as one masters the initial fear of parachuting from an aircraft by doing it over and over again. Eventually, it is difficult to remember how frightening the event was initially. In giving first aid to the emotionally disturbed service member, you should let him follow this natural pattern. Encourage him to talk. Be a good listener. Let him tell, in his own words, what actually happened. If home front problems or worries have contributed to the stress, it will help him to talk about them. Your patient listening will prove to him that you are interested in him, and by describing his personal problem, he can work at mastering his fear. If he becomes overwhelmed in the telling, suggest a cup of coffee or a break. Whatever you do, assure him that you will listen again as soon as he is ready. Do try to help put the service member's perception of what happened back into realistic perspective; but DO NOT argue about it.

b. *Activity*.

(1) A person who is emotionally disturbed as the result of a combat action is a casualty of anxiety and fear. He is disabled because he has become temporarily overwhelmed by his anxiety. A good way to control fear is through activity. Almost all service members, for example, experience a considerable sense of anxiety and fear while they are poised, awaiting the opening of a big offensive; but this is normally relieved, and they actually feel better once they begin to move into action. They take pride in effective performance and pleasure in knowing that they are good service members, perhaps being completely unaware that overcoming their initial fear was their first major accomplishment.

(2) Useful activity is very beneficial to the emotionally disturbed service member who is not physically incapacitated. After you help a service member get over his initial fear, help him to regain some self-confidence. Make him realize his job is continuing by finding him something useful to do. Encourage him to be active. Get him to help load trucks, clean up debris, or dig fighting positions. If possible, get him back to his usual duty. Seek out his strong points and help him apply them. Avoid having him just sit around. You may have to provide direction by telling him what to do and where to do it. The instructions should be clear and simple and should be repeated. A person who has panicked is likely to argue. Respect his feelings, but point out more immediate, obtainable, and demanding needs. Channel his excessive energy and, above all, DO NOT argue. If you cannot get him interested in doing more profitable work, it may be necessary to enlist aid in controlling his overactivity before it spreads to the group and

results in more panic. Prevent the spread of such infectious feelings by restraining and segregating if necessary.

(3) Involvement in activity helps a service member in three ways; he—

- Forgets himself.
- Has an outlet for his excessive tensions.
- Proves to himself he can do something useful.

c. Rest. There are times, particularly in combat, when physical exhaustion is a principal cause for emotional reactions. A unit sleep plan should be established and implemented. When possible, service members should be given a safe and relatively comfortable area in which to sleep. Examples would be an area away from heavy traffic, noise, and congestion or a place that is clean and dry and protected from environmental conditions. The more uninterrupted sleep a service member gets the better he will be able to function in the tactical environment.

d. Hygiene. Field hygiene is an important ingredient in a service member's morale. A service member who is dirty and unkempt will not function as well as a service member who has had the opportunity to bathe and put on clean, dry clothing. During combat, unit leaders should stress the importance of personal hygiene. Good personal hygiene not only improves morale, it also is a preventive measure against disease and nonbattle injury (DNBI).

e. Group Activity. You have probably already noticed that a person works, faces danger, and handles serious problems better if he is a member of a closely-knit group. Each service member in the team supports the other team members. Esprit de corps is built because the service members have the same interests, goals, and mission, and as a result they are more productive; furthermore, they are less worried because everyone is involved. It is this spirit that takes a strategic hill in battle. It is so powerful that it is one of the most effective tools you have in your *psychological first aid bag*. Getting the service member back into the team or squad activities will reestablish his sense of belonging and security and will go far toward making him a useful member of the unit.

8-12. Reactions and Limitations

Up to this point the discussion has been primarily about the feelings of the emotionally distressed service member. What about your feelings toward him? Whatever the situation, you will have emotional reactions (conscious or unconscious) toward this service member. Your reactions can either help or hinder your ability to help him. When you are tired or worried, you may very easily become impatient with him if he is unusually slow or exaggerates. You may even feel resentful toward him. At times when many physically wounded lie about you, it will be especially natural for you to resent disabilities that you cannot see. Physical wounds can be seen and easily accepted. Emotional reactions are more difficult to accept as injuries. On the other hand, will you tend to be overly sympathetic? Excessive sympathy for an incapacitated person can be as harmful as negative feelings in your relationship with him. He needs strong help, but not your sorrow. To overwhelm him with pity will make him feel even more inadequate. You must expect your buddy to recover, to be able to return to duty, and to become a useful service member again. This expectation should be displayed in your behavior and attitude as well as in what you say. If he can see your calmness, confidence, and competence, he will be reassured and will feel a sense of greater security.

8-13. Stress Reactions

See Tables 8-1, 8-2, and 8-3 for more information.

Table 8-1. Mild Stress Reaction

PHYSICAL SIGNS*	EMOTIONAL SIGNS*
1. TREMBLING, TEARFUL 2. JUMPINESS, NERVOUSNESS 3. COLD SWEAT, DRY MOUTH 4. POUNDING HEART, DIZZINESS 5. INSOMNIA, NIGHTMARES 6. NAUSEA, VOMITING, DIARRHEA 7. FATIGUE 8. THOUSAND-YARD STARE 9. DIFFICULTY THINKING, SPEAKING, AND COMMUNICATING	1. ANXIETY, INDECISIVENESS 2. IRRITABLE, COMPLAINING 3. FORGETFUL, UNABLE TO CONCENTRATE 4. EASILY STARTLED BY NOISE, MOVEMENT 5. GRIEF, TEARFUL 6. ANGER, BEGINNING TO LOSE CONFIDENCE IN SELF AND UNIT

SELF- AND BUDDY AID
1. CONTINUE MISSION PERFORMANCE, FOCUS ON IMMEDIATE MISSION. 2. EXPECT SERVICE MEMBER TO PERFORM ASSIGNED DUTIES. 3. REMAIN CALM AT ALL TIMES; BE DIRECTIVE AND IN CONTROL. 4. LET SERVICE MEMBER KNOW HIS REACTION IS NORMAL, AND THAT THERE IS NOTHING SERIOUSLY WRONG WITH HIM. 5. KEEP SERVICE MEMBER INFORMED OF THE SITUATION, OBJECTIVES, EXPECTATIONS, AND SUPPORT. CONTROL RUMORS. 6. BUILD SERVICE MEMBER'S CONFIDENCE, TALK ABOUT SUCCEEDING. 7. KEEP SERVICE MEMBER PRODUCTIVE (WHEN NOT RESTING) THROUGH RECREATIONAL ACTIVITIES, EQUIPMENT MAINTENANCE.

8. ENSURE SERVICE MEMBER MAINTAINS GOOD PERSONAL HYGIENE.
9. ENSURE SERVICE MEMBER EATS, DRINKS, AND SLEEPS AS SOON AS POSSIBLE.
10. LET SERVICE MEMBER TALK ABOUT HIS FEELINGS. DO NOT "PUT DOWN" HIS FEELINGS OF GRIEF OR WORRY. GIVE PRACTICAL ADVICE AND PUT EMOTIONS INTO PERSPECTIVE.

* MOST OR ALL OF THESE SIGNS ARE PRESENT IN MILD STRESS REACTION. THEY CAN BE PRESENT IN ANY NORMAL SERVICE MEMBER IN COMBAT YET HE CAN STILL DO HIS JOB.

Table 8-2. More Serious Stress Reaction

PHYSICAL SIGNS*	EMOTIONAL SIGNS*
1. CONSTANTLY MOVES AROUND 2. FLINCHING OR DUCKING AT SUDDEN SOUNDS 3. SHAKING, TREMBLING (WHOLE BODY OR ARMS) 4. CANNOT USE PART OF BODY, NO PHYSICAL REASON (HAND, ARM, LEGS) 5. CANNOT SEE, HEAR, OR FEEL (PARTIAL OR COMPLETE LOSS) 6. PHYSICAL EXHAUSTION, CRYING 7. FREEZING UNDER FIRE, OR TOTAL IMMOBILITY 8. VACANT STARES, STAGGERS, SWAYS WHEN STANDS 9. PANIC RUNNING UNDER FIRE	1. RAPID AND/OR INAPPROPRIATE TALKING 2. ARGUMENTATIVE, RECKLESS MOVEMENTS/ACTIONS 3. INATTENTIVE TO PERSONAL HYGIENE 4. INDIFFERENT TO DANGER 5. MEMORY LOSS 6. SEVERE STUTTERING, MUMBLING, OR CANNOT SPEAK AT ALL 7. INSOMNIA, NIGHTMARES 8. SEEING OR HEARING THINGS THAT DO NOT EXIST 9. RAPID EMOTIONAL SHIFTS 10. SOCIAL WITHDRAWAL 11. APATHETIC 12. HYSTERICAL OUTBURSTS 13. FRANTIC OR STRANGE BEHAVIOR

TREATMENT PROCEDURES**

1. IF A SERVICE MEMBER'S BEHAVIOR ENDANGERS THE MISSION, SELF, OR OTHERS, DO WHATEVER IS NECESSARY TO CONTROL HIM.
2. IF THE SERVICE MEMBER IS UPSET, CALMLY TALK HIM INTO COOPERATING.
3. IF CONCERNED ABOUT THE SERVICE MEMBER'S RELIABILITY:
   - UNLOAD HIS WEAPON.
   - TAKE WEAPON IF SERIOUSLY CONCERNED.
   - PHYSICALLY RESTRAIN HIM ONLY WHEN NECESSARY FOR SAFETY OR TRANSPORTATION.
4. REASSURE EVERYONE THAT THE SIGNS ARE PROBABLY JUST STRESS REACTION AND WILL QUICKLY IMPROVE.
5. IF STRESS REACTION SIGNS CONTINUE:
   - GET THE SERVICE MEMBER TO A SAFER PLACE.
   - DO NOT LEAVE THE SERVICE MEMBER ALONE, KEEP SOMEONE HE KNOWS WITH HIM.
   - NOTIFY SENIOR NONCOMMISSIONED OFFICER (NCO) OR OFFICER.
   - HAVE THE SERVICE MEMBER EXAMINED BY MEDICAL PERSONNEL.

Table 8-2. More Serious Stress Reaction (Continued)

TREATMENT PROCEDURES**

6. GIVE THE SERVICE MEMBER EASY TASKS TO DO WHEN NOT SLEEPING, EATING, OR RESTING.
7. ASSURE THE SERVICE MEMBER HE WILL RETURN TO FULL DUTY IN 24 HOURS; AND, RETURN HIM TO NORMAL DUTIES AS SOON AS HE IS READY.

* THESE SIGNS ARE PRESENT IN ADDITION TO THE SIGNS OF MILD STRESS REACTION.

** DO THESE PROCEDURES IN ADDITION TO THE SELF- AND BUDDY AID CARE.

Table 8-3. Preventive Measures to Combat Stress Reaction

1. WELCOME NEW MEMBERS INTO YOUR TEAM, GET TO KNOW THEM QUICKLY. IF YOU ARE NEW, BE ACTIVE IN MAKING FRIENDS.
2. BE PHYSICALLY FIT (STRENGTH, ENDURANCE, AND AGILITY).
3. KNOW AND PRACTICE LIFESAVING SELF- AND BUDDY AID.
4. PRACTICE RAPID RELAXATION TECHNIQUES (FM 22-51).
5. HELP EACH OTHER OUT WHEN THINGS ARE TOUGH AT HOME OR IN THE UNIT.
6. KEEP INFORMED; ASK YOUR LEADER QUESTIONS, IGNORE RUMORS.
7. WORK TOGETHER TO GIVE EVERYONE FOOD, WATER, SHELTER, HYGIENE, AND SANITATION.
8. SLEEP WHEN MISSION AND SAFETY PERMIT; LET EVERYONE GET TIME TO SLEEP.
   - SLEEP ONLY IN SAFE PLACES AND BY STANDING OPERATING PROCEDURE (SOP).
   - IF POSSIBLE, SLEEP 6 TO 9 HOURS PER DAY.
   - TRY TO GET AT LEAST 4 HOURS SLEEP PER DAY.
   - GET GOOD SLEEP BEFORE GOING ON SUSTAINED OPERATIONS.
   - CATNAP WHEN YOU CAN, BUT ALLOW TIME TO WAKE UP FULLY.
   - CATCH UP ON SLEEP AFTER GOING WITHOUT.

APPENDIX A

FIRST AID CASE AND KITS, DRESSINGS, AND BANDAGES

A-1. First Aid Case with Field Dressings and Bandages

Every service member is issued a first aid case (Figure A-1A) with a field first aid dressing encased in a plastic wrapper (Figure A-1B). He carries it at all times for his use. The field first aid dressing is a standard sterile (germ-free) compress or pad with bandages attached (Figure A-1C). This dressing is used to cover the wound, to protect against further contamination, and to stop bleeding (pressure dressing). When a service member administers first aid to another person, he must remember to use the wounded person's dressing; he may need his own later. The service member must check his first aid case regularly and replace any used or missing dressing. The field first aid dressing may normally be obtained from his unit supply.

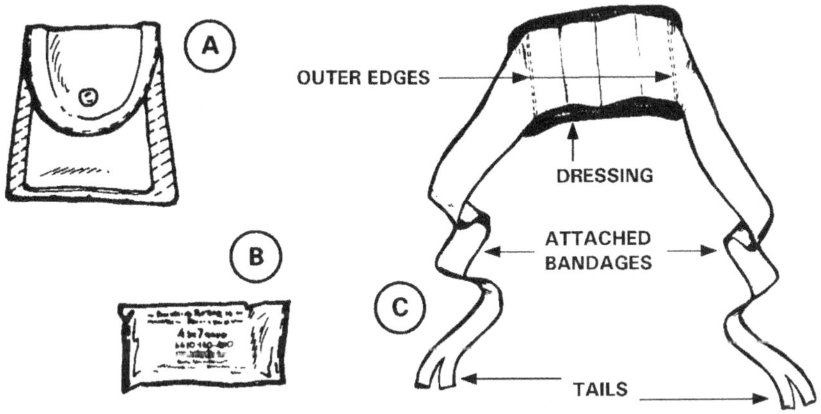

Figure A-1. Field first aid case and dressing (Illustrated A—C).

A-2. General Purpose First Aid Kits

General-purpose first aid kits are listed in the common table of allowances (CTA) 8-100. The operators, crew, and passengers carry these kits on Army vehicles, aircraft, and boats for use. Individuals designated by unit TSOP to be responsible for these kits are required to check them regularly and replace all items used. The general-purpose kit and its contents can be obtained through the unit supply system.

NOTE

Periodically check the dressings (for holes or tears in the packaging) and the medicines (for expiration date) that are in the first aid kits. If necessary, replace defective or outdated items.

A-3. Dressings

Dressings are sterile pads or compresses used to cover wounds. They usually are made of gauze or cotton wrapped in gauze (Figure A-1C). In addition to the standard field first aid dressing, other dressings such as sterile gauze compresses and small sterile compresses on adhesive strips may be available under CTA 8-100.

A-4. Standard Bandages

a. Standard bandages are made of gauze or muslin and are used over a sterile dressing to secure the dressing in place, to close off its edge from dirt and germs, and to create pressure on the wound and control bleeding. A bandage can also support an injured part or secure a splint.

b. Tailed bandages may be attached to the dressing as indicated on the field first aid dressing (Figure A-1C).

A-5. Triangular and Cravat (Swathe) Bandages

a. Triangular and cravat (or swathe) bandages (Figure A-2) are fashioned from a triangular piece of muslin (37 by 37 by 52 inches) provided in the general-purpose first aid kit. If it is folded into a strip, it is called a cravat. Two safety pins are packaged with each bandage. These bandages are valuable in an emergency since they are easily applied.

b. To improvise a triangular bandage, cut a square of available material, slightly larger than 3 feet by 3 feet, and *fold it diagonally*. If two bandages are needed, cut the material along the diagonal fold.

c. A cravat can be improvised from such common items as T-shirts, other shirts, bed linens, trouser legs, scarfs, or any other item made of pliable and durable material that can be folded, torn, or cut to the desired size.

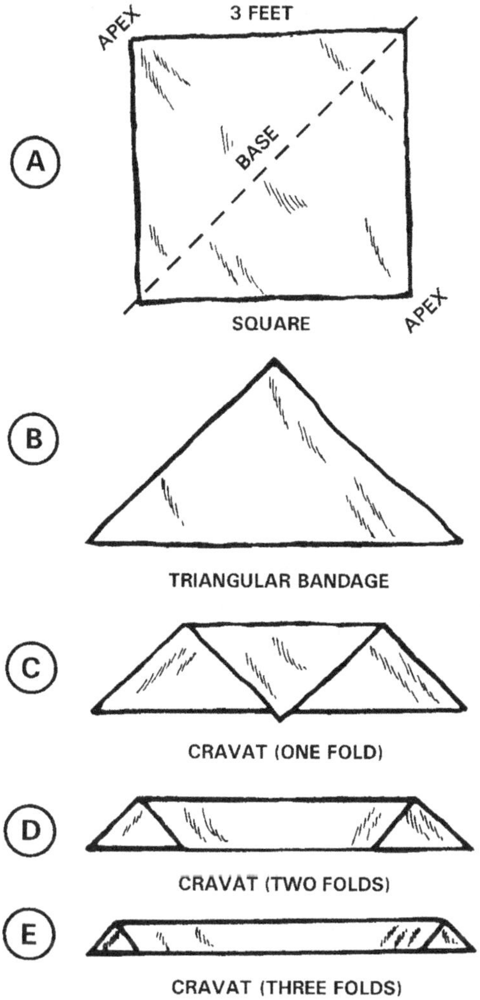

Figure A-2. Triangular and cravat bandages (Illustrated A—E).

APPENDIX B

RESCUE AND TRANSPORTATION PROCEDURES

B-1. General

A basic principle of first aid is to evaluate the casualty's injuries and administer first aid before moving him. However, adverse situations or conditions may jeopardize the lives of both the rescuer and the casualty if this is done. It may be necessary *first to rescue* the casualty before first aid can be effectively or safely given. The life and/or the well-being of the casualty will depend as much upon the manner in which he is *rescued and transported,* as it will upon the first aid and medical treatment he receives. Rescue actions must be done quickly and safely. Careless or rough handling of the casualty during rescue operations can aggravate his injuries.

B-2. Principles of Rescue Operations

a. When faced with the necessity of rescuing a casualty who is threatened by hostile action, fire, water, or any other immediate hazard, DO NOT take action without first determining the extent of the hazard and your ability to handle the situation. DO NOT become a casualty.

b. The rescuer must evaluate the situation and analyze the factors involved. This evaluation involves three major steps:

- Identify the task.
- Evaluate circumstances of the rescue.
- Plan the action.

B-3. Considerations

a. First determine if a rescue attempt is actually needed. It is a waste of time, equipment, and personnel to rescue someone not in need of rescuing. It is also a waste to look for someone who is not lost or needlessly risk the lives of the rescuer(s). In planning a rescue, attempt to obtain the following information:

- Who, what, where, when, why, and how the situation happened?

- How many casualties are involved and the nature of their injuries?

- What is the tactical situation?

- What are the terrain features and the location of the casualties?

- Will there be adequate assistance available to aid in the rescue/evacuation?

- Can first aid and/or medical treatment be provided at the scene; will the casualties require movement to a safer location?

- What specialized equipment will be required for the rescue operation?

- Is the rescue area contaminated? Will decontamination equipment and materiel be required for casualties, rescue personnel, and rescue equipment?

- How much time is available?

b. The time element can play a significant role in how the rescue is attempted. If the casualties are in imminent danger of losing their lives (such as near a burning vehicle or in a burning building) the time available will be relatively short and will sometimes cause a rescuer to compromise planning stages and/or the first aid which can be given. However, if the casualty is in a relatively secure area and his physical condition is strong, more deliberate planning can take place. A realistic estimate of time available must be made as quickly as possible to determine action time remaining. The key elements are the casualty's physical and mental condition, the tactical situation, and the environment.

B-4. Plan of Action

a. The casualty's ability to endure is of primary importance in estimating the time available. Age, physical condition, and extent of wounds and/or injuries will differ from casualty to casualty. Therefore, to determine the time available, you will have to consider—

- Endurance time of the casualty.

- Extent of injuries.

- Type of situation.
- Personnel and/or equipment availability.
- Weather.
- Terrain (natural and man-made).
- Environment (contaminated or uncontaminated).

b. In respect to terrain, you must consider altitude and visibility. In some cases, the casualty may be of assistance because he knows more about the particular terrain or situation than you do. Maximum use of secure/reliable trails or roads is essential.

c. When taking weather into account, ensure that blankets and/or rain gear are available. Even a mild rain can complicate a normally simple rescue. In high altitudes and/or extreme cold and gusting winds, the time available is critically shortened. Be prepared to provide shelter and warmth for the casualty as well as the rescuers.

B-5. Proper Handling of Casualties

a. You may have saved the casualty's life through the application of appropriate first aid measures. However, his life can be lost through rough handling or careless transportation procedures. Before you attempt to move the casualty—

- Evaluate the type and extent of his injuries.
- Ensure that dressings over wounds are adequately reinforced.
- Ensure that fractured bones are properly immobilized and supported to prevent them from cutting through muscle, blood vessels, and skin.

b. Based upon your evaluation of the type and extent of the casualty's injury and your knowledge of the various manual carries, you must select the best possible method of manual transportation. If the casualty is conscious, tell him how he is to be transported. This will help allay his fear of movement and gain his cooperation and confidence.

c. Buddy aid for chemical agent casualties includes those actions required to prevent an incapacitated casualty from receiving additional injury from the effects of chemical hazards. If a casualty is physically unable to decontaminate himself or administer the proper chemical agent antidote, the casualty's buddy assists him and assumes responsibility for his care. Buddy-aid includes—

- Administering the proper chemical agent antidote.
- Decontaminating the incapacitated casualty's exposed skin.
- Ensuring that his protective ensemble remains correctly emplaced.
- Maintaining respiration.
- Controlling bleeding.
- Providing other standard first aid measures
- Transporting the casualty out of the contaminated area.

B-6. Positioning the Casualty

The first step in any manual carry is to position the casualty to be lifted. If he is conscious, he should be told how he is to be positioned and transported. This helps lessen his fear of movement and to gain his cooperation. It may be necessary to roll the casualty onto his abdomen, or his back, depending upon the position in which he is lying and the particular carry to be used.

a. To roll a casualty onto his abdomen, kneel at the casualty's uninjured side.

(1) Place his arms above his head; cross his ankle which is farther from you over the one that is closer to you.

(2) Place your hands on the shoulder which is farther from you; place your other hand in the area of his hip or thigh (Figure B-1).

(3) Roll him gently toward you onto his abdomen (Figure B-2).

b. To roll a casualty onto his back, follow the same procedure described in *a* above, except gently roll the casualty onto his back, rather than onto his abdomen.

Figure B-1. Positioning the casualty.

Figure B-2. Rolling casualty onto his abdomen.

B-7. Medical Evacuation and Transportation of Casualties

a. Medical evacuation of the sick and wounded (with en route medical care) is the responsibility of medical personnel who have been provided special training and equipment. Therefore, unless a good reason for you to transport a casualty arises, wait for some means of medical evacuation to be provided. When the situation is urgent and you are unable to obtain medical assistance or know that no medical evacuation assets are available, you will have to transport the casualty. For this reason, you must know how to transport him without increasing the seriousness of his condition.

b. Transporting a casualty by litter (FM 8-10-6) is safer and more comfortable for him than by manual means; it is also easier for you.

Manual transportation, however, may be the only feasible method because of the terrain or the combat situation; or it may be necessary to save a life. In these situations, the casualty should be transferred to a litter as soon as one can be made available or improvised.

B-8. Manual Carries

Casualties carried by manual means must be carefully and correctly handled, otherwise their injuries may become more serious or possibly fatal. Situation permitting, transport of a casualty should be organized and unhurried. Each movement should be performed as deliberately and gently as possible. Casualties should not be moved before the type and extent of injuries are evaluated and the required first aid is administered. The exception to this occurs when the situation dictates immediate movement for safety purposes (for example, it may be necessary to remove a casualty from a burning vehicle); that is, the situation dictates that the urgency of casualty movement outweighs the need to administer first aid. Manual carries are tiring for the bearers and involve the risk of increasing the severity of the casualty's injury. In some instances, however, they are essential to save the casualty's life. Although manual carries are accomplished by one or two bearers, the two-man carries are used whenever possible. They provide more comfort to the casualty, are less likely to aggravate his injuries, and are also less tiring for the bearers. The distance a casualty can be carried depends on many factors, such as—

- Nature of the casualty's injuries.
- Strength and endurance of the bearer(s).
- Weight of the casualty.
- Obstacles encountered during transport (natural or manmade).
- Type of terrain.

a. One-man Carries. These carries should be used when only one bearer is available to transport the casualty.

(1) The *fireman's carry* (Figure B-3) is one of the easiest ways for one individual to carry another. After an unconscious or disabled casualty has been properly positioned, he is raised from the ground, then supported and placed in the carrying position.

(*a*) After rolling the casualty onto his abdomen, straddle him. Extend your hands under his chest and lock them together.

(*b*) Lift the casualty to his knees as you move backward.

(*c*) Continue to move backward, thus straightening the casualty's legs and locking his knees.

(*d*) Walk forward, bringing the casualty to a standing position; tilt him slightly backward to prevent his knees from buckling.

(*e*) As you maintain constant support of the casualty with one arm, free your other arm, quickly grasp his wrist, and raise his arm high. Instantly pass your head under his raised arm, releasing it as you pass under it.

(*f*) Move swiftly to face the casualty and secure your arms around his waist. Immediately place your foot between his feet and spread them apart (approximately 6 to 8 inches).

(*g*) Grasp the casualty's wrist and raise his arm high over your head.

(*h*) Bend down and pull the casualty's arm over and down on your shoulder, bringing his body across your shoulders. At the same time, pass your arm between his legs.

(*i*) Grasp the casualty's wrist with one hand, and place your other hand on your knee for support.

(*j*) Rise with the casualty positioned correctly. Your other hand is free for use.

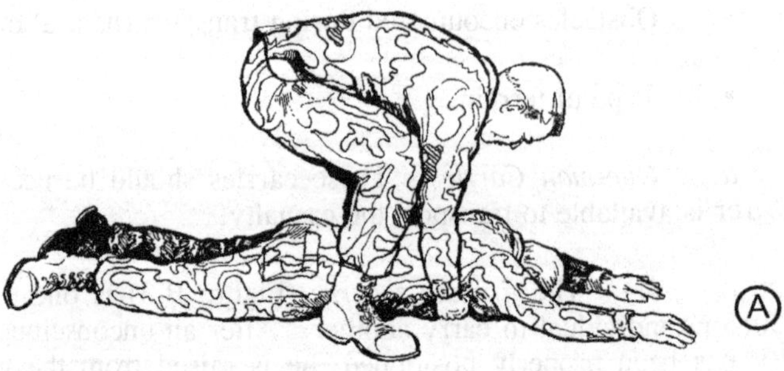

Figure B-3. Fireman's carry (Illustrated A—J).

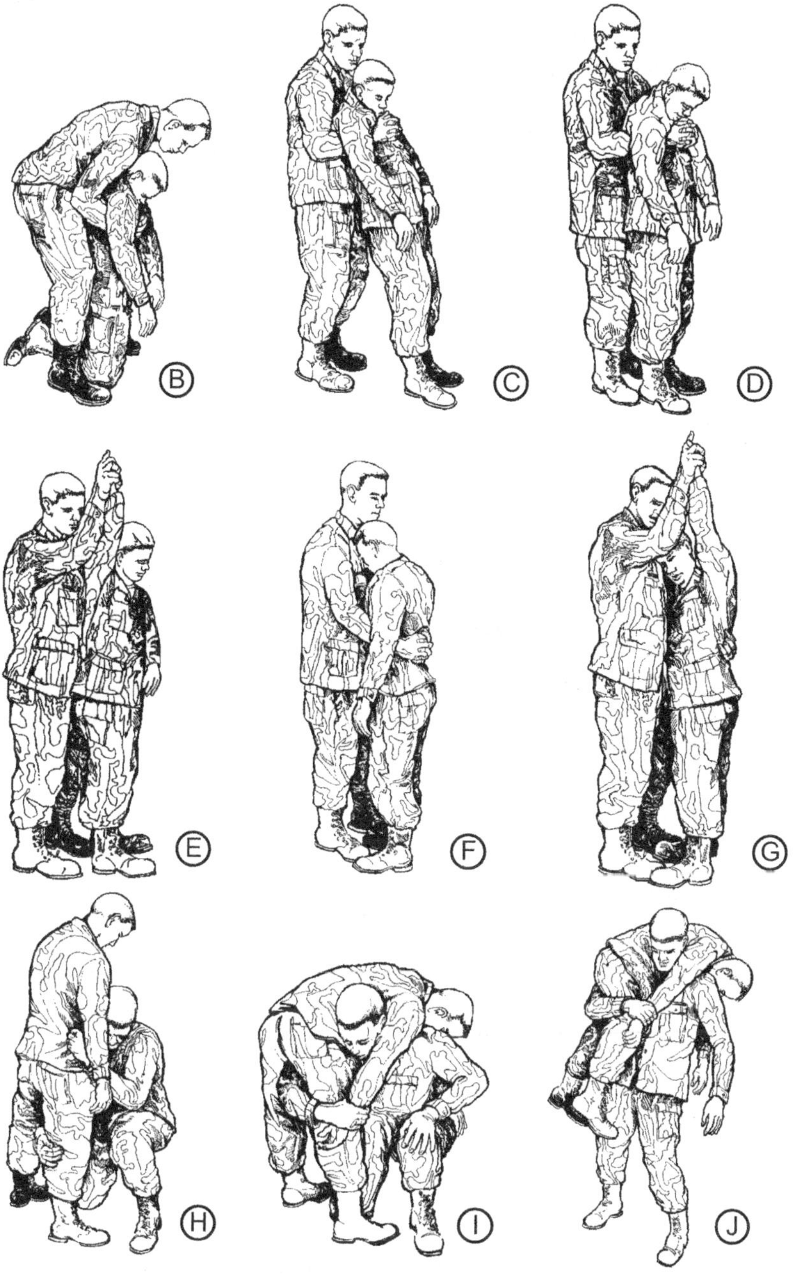

Figure B-3. Fireman's carry (Illustrated A—J) (Continued).

(2) The alternate method of the *fireman's carry* for raising a casualty from the ground is illustrated in Figure B-4; however, it should be used only when the bearer believes it to be safer for the casualty because of the location of his wounds. When the alternate method is used, care must be taken to prevent the casualty's head from snapping back and causing a neck injury. The steps for raising a casualty from the ground for the fireman's carry are also used in other one-man carries.

(*a*) Kneel on one knee at the casualty's head and face his feet. Extend your hands under his armpits, down his sides, and across his back.

(*b*) As you rise, lift the casualty to his knees. Then secure a lower hold and raise him to a standing position with his knees locked.

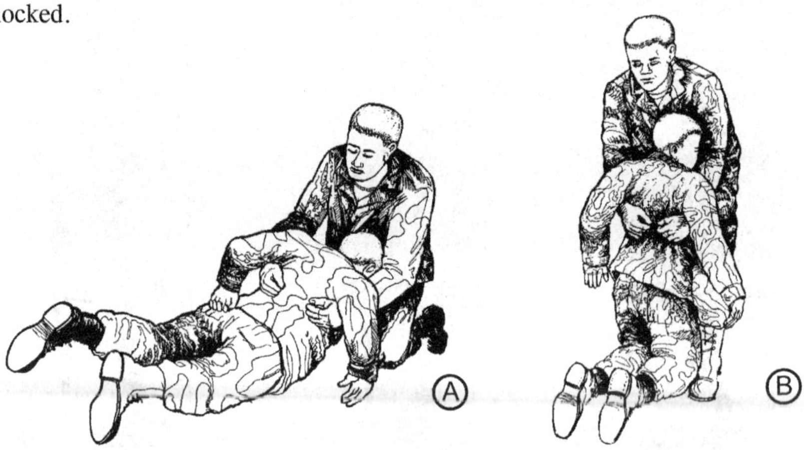

Figure B-4. Fireman's carry (alternate method) for lifting a casualty to a standing position (Illustrated A—B).

(3) In the *supporting carry* (Figure B-5), the casualty must be able to walk or at least hop on one leg, using the bearer as a crutch. This carry can be used to assist him as far as he is able to walk or hop.

(*a*) Raise the casualty from the ground to a standing position by using the fireman's carry.

(*b*) Grasp the casualty's wrist and draw his arm around your neck.

(*c*) Place your arm around his waist. The casualty is now able to walk or hop using you as a support.

Figure B-5. Supporting carry.

(4) The *arms carry* (Figure B-6) is useful in carrying a casualty for a short distance (up to 50 meters) and for placing him on a litter.

(*a*) Raise or lift the casualty from the ground to a standing position, as in the fireman's carry.

(*b*) Place one arm under the casualty's knees and your other arm around his back.

(*c*) Lift the casualty.

(*d*) Carry the casualty high to lessen fatigue.

Figure B-6. Arms carry.

(5) Only a conscious casualty can be transported by the *saddleback carry* (Figure B-7), because he must be able to hold onto the bearer's neck. To use this technique–

(*a*) Raise the casualty to an upright position, as in the fireman's carry.

(*b*) Support the casualty by placing an arm around his waist. Move to the casualty's side. Have the casualty put his arm around your neck and move in front of him with your back to support him.

(*c*) Have the casualty encircle his arms around your neck

(*d*) Stoop, raise him on your back and clasp your hands together beneath his thighs, if possible.

Figure B-7. Saddleback carry.

(6) In the *pack-strap carry* (Figure B-8), the casualty's weight rests high on the your back. This makes it easier for you to carry the casualty a moderate distance (50 to 300 meters). To eliminate the possibility of injury to the casualty's arms, you must hold his arms in a palms-down position.

(*a*) Lift the casualty from the ground to a standing position, as in the fireman's carry.

(*b*) Support the casualty with your arms around him and grasp his wrist closer to you.

(*c*) Place his arm over your head and across your shoulders.

(*d*) Move in front of him while still supporting his weight against your back.

(*e*) Grasp his other wrist and place this arm over your shoulder.

(*f*) Bend forward and raise or hoist the casualty as high on your back as possible so that his weight is resting on your back.

NOTE

Once the casualty is positioned on the bearer's back, the bearer remains as erect as possible to prevent straining or injuring his back.

Figure B-8. Pack-strap carry.

(7) The *pistol-belt carry* (Figure B-9) is the best one-man carry for a long distance (over 300 meters). The casualty is securely supported upon your shoulders by a belt. Both your hands and the casualty's (if conscious) are free for carrying a weapon or equipment, or climbing obstacles. With your hands free and the casualty secured in place, you are also able to creep through shrubs and under low-hanging branches.

(*a*) Link two pistol belts (or three, if necessary) together to form a sling. Place the sling under the casualty's thighs and lower back so that a loop extends from each side.

NOTE

If pistol belts are not available for use, other items such as a rifle sling, two cravat bandages, two litter straps, or any other suitable material, which will not cut or bind the casualty may be used.

(*b*) Lie face up between the casualty's outstretched legs. Thrust your arms through the loops and grasp his hands and trouser leg on his injured side.

(*c*) Roll toward the casualty's uninjured side onto your abdomen, bringing him onto your back. Adjust the sling, if necessary.

(*d*) Rise to a kneeling position. The belt will hold the casualty in place.

(*e*) Place one hand on your knee for support and rise to an upright position. (The casualty is supported on your shoulders.)

(*f*) Carry the casualty with your hands free for use in rifle firing, climbing, or surmounting obstacles.

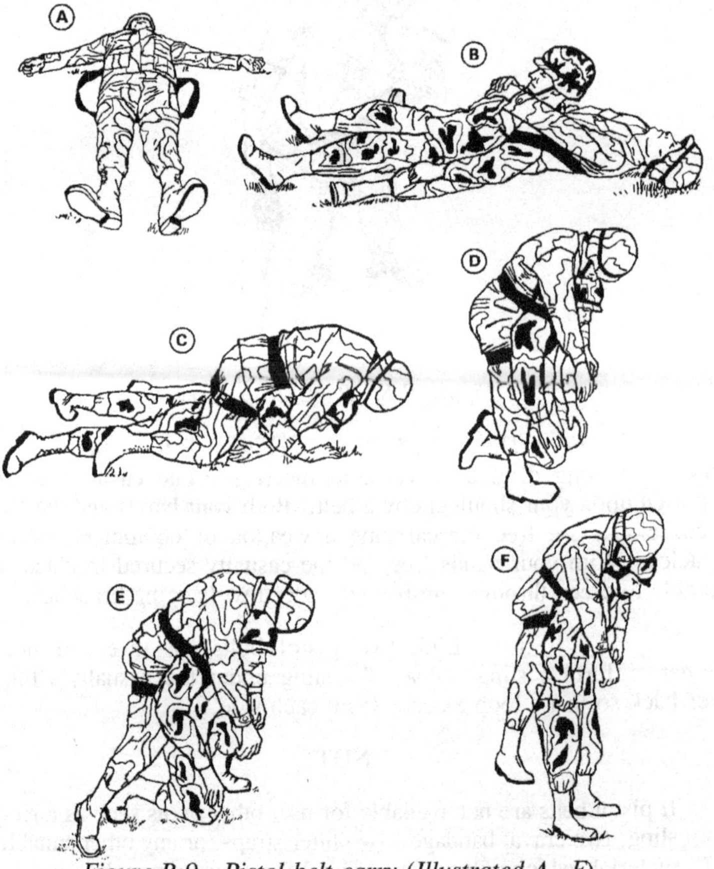

Figure B-9. Pistol-belt carry (Illustrated A—F).

(8) The *pistol-belt drag* (Figure B-10), as well as other drags, is generally used for short distances (up to 50 meters). This drag is useful in combat, since both the bearer and the casualty can remain closer to the ground than in any other drags.

(*a*) Extend two pistol belts or similar objects to their full length and join them together to make a continuous loop.

(*b*) Roll the casualty onto his back, as in the fireman's carry.

(*c*) Pass the loop over the casualty's head, and position it across his chest and under his armpits. Then cross the remaining portion of the loop, thus forming a figure eight. Keep tension on the belts so they do not come unhooked.

(*d*) Lie on your side facing the casualty.

(*e*) Slip the loop over your head and turn onto your abdomen. This enables you to drag the casualty as you crawl.

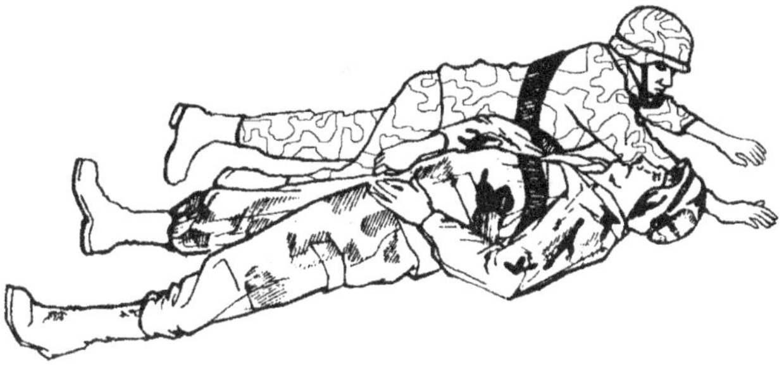

Figure B-10. Pistol-belt drag.

(9) The *neck drag* (Figure B-11) is useful in combat because the bearer can transport the casualty as he creeps behind a low wall or shrubbery, under a vehicle, or through a culvert. If the casualty is unconscious, his head must be protected from the ground. The neck drag cannot be used if the casualty has a broken arm.

NOTE

If the casualty is conscious, he may clasp his hands together around your neck.

(*a*) Tie the casualty's hands together at the wrists.

(*b*) Straddle the casualty in a kneeling face-to-face position.

(*c*) Loop the casualty's tied hands over and around your neck.

(*d*) Crawl forward dragging the casualty with you.

NOTE

If the casualty is unconscious, protect his head from the ground.

Figure B-11. Neck drag.

(10) The *cradle drop drag* (Figure B-12) is effective in moving a casualty up or down steps.

(*a*) Kneel at the casualty's head (with him lying on his back). Slide your hands, with palms up, under the casualty's shoulders and get a firm hold under his armpits.

(*b*) Rise (partially), supporting the casualty's head on one of your forearms. (You may bring your elbows together and let the casualty's head rest on both of your forearms.)

(*c*) Rise and drag the casualty backward. (The casualty is in a semisitting position.)

(*d*) Back down the steps, supporting the casualty's head and body and letting his hips and legs drop from step to step.

NOTE

If the casualty needs to be moved up the steps, you should back up the steps, using the same procedure.

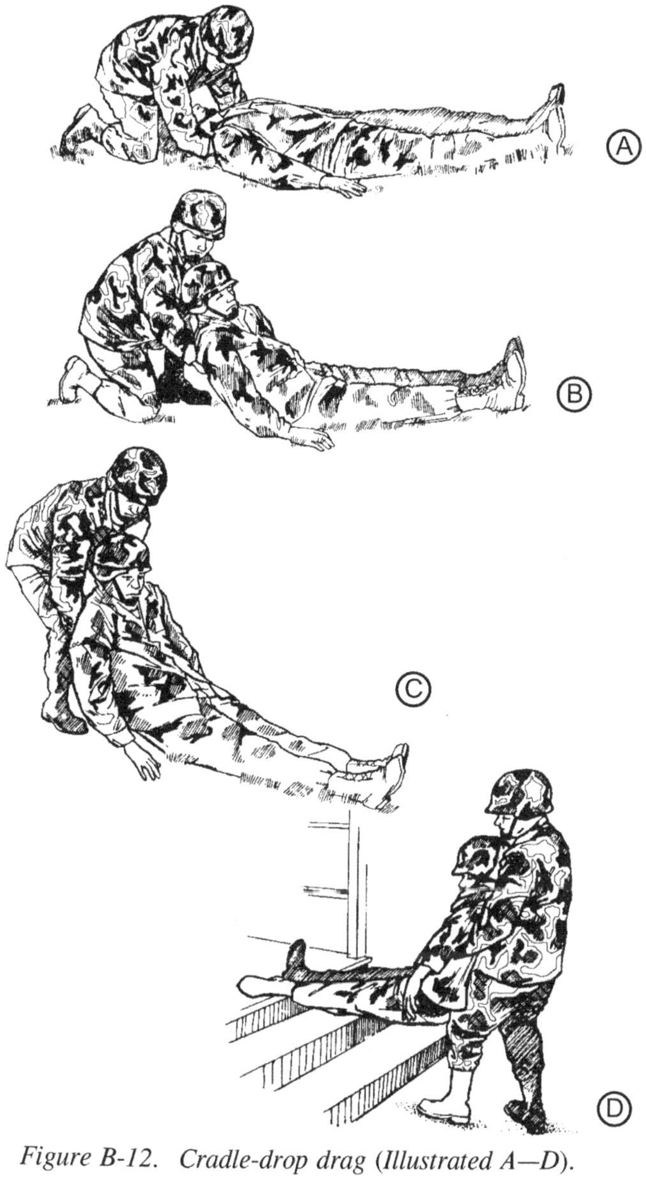

Figure B-12. Cradle-drop drag (Illustrated A—D).

(11) The *LBE carry using the bearer's LBE* can be used with a conscious casualty (Figure B-13).

(*a*) Loosen all suspenders on your LBE.

(*b*) Have the casualty place one leg into the loop formed by your suspenders and pistol belt.

(*c*) Squat in front of the standing casualty. Have him place his other leg into the loop, also.

(*d*) Have the casualty place his arms over your shoulders, lean forward onto your back, and lock his hands together.

(*e*) Stand up and lean forward into a comfortable position.

(*f*) Continue the mission.

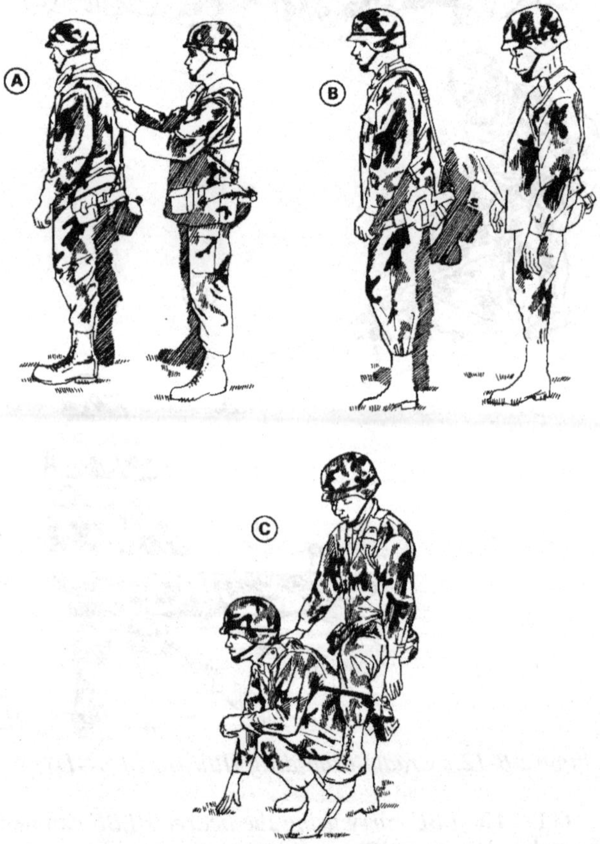

Figure B-13. Load bearing equipment carry using bearer's LBE (conscious casualty) (Illustrated A—F).

Figure B-13. Load bearing equipment carry using bearer's LBE (conscious casualty) (Illustrated A—F) (Continued).

(12) The *LBE carry using the bearer's LBE* can be used with an unconscious casualty or one who cannot stand (Figure B-14).

(*a*) Position the casualty on the flat of his back.

(*b*) Remove your LBE and loosen all suspender straps.

(*c*) Lift the casualty's leg and place it through the loop formed by your suspenders and pistol belt. Then place the other leg through the same loop. The LBE is moved up until the pistol belt is behind the casualty's thighs.

(*d*) Lay between the casualty's legs; work your arms through the LBE suspenders.

(*e*) Grasp the casualty's hand (on the injured side), and roll the casualty (on his uninjured side) onto your back.

(*f*) Rise to one knee and then push into a standing position.

(*g*) Bring the casualty's arms over your shoulders. Grasp his hands and secure them if the casualty is unconscious. If the casualty is conscious, have him lock his hands in front if he is able to do so.

(*h*) Lean forward into a comfortable position and continue the mission.

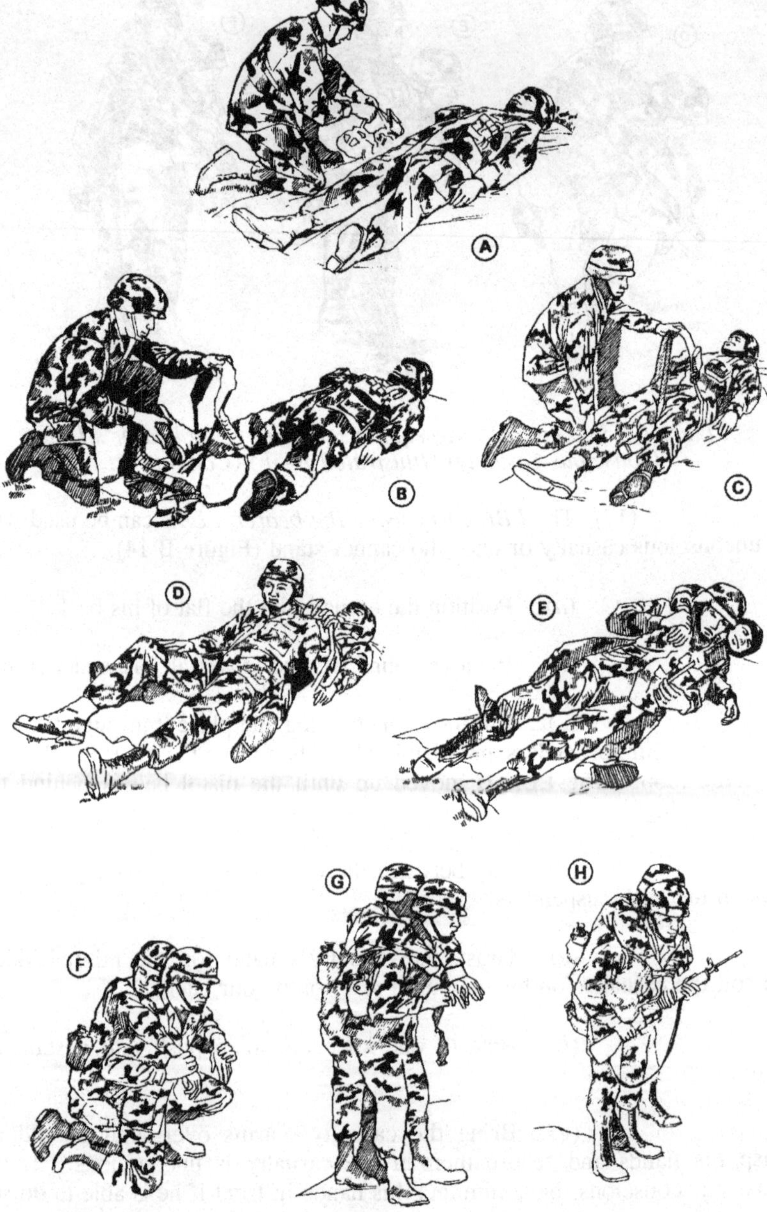

Figure B-14. Load bearing equipment carry using bearer's LBE (unconscious casualty or one that cannot stand) (Illustrated A—H).

(13) The *LBE carry using the casualty's LBE* (Figure B-15) can be used with a conscious or unconscious casualty.

(*a*) Position the casualty on his back with his LBE on.

(*b*) Loosen the casualty's two front suspenders.

(*c*) Position yourself between the casualty's legs, and slip your arms into the casualty's two front suspenders (up to your shoulders).

(*d*) Work his arms out of his LBE suspenders.

(*e*) Grasp the casualty's hand (on the injured side), and roll him (on his uninjured side) onto your back.

(*f*) Rise to one knee, then into a standing position.

(*g*) Grasp the casualty's hands and secure them, if the casualty is unconscious. Have the casualty lock his hands in front of you, if he is conscious.

(*h*) Lean forward into a comfortable position and continue the mission.

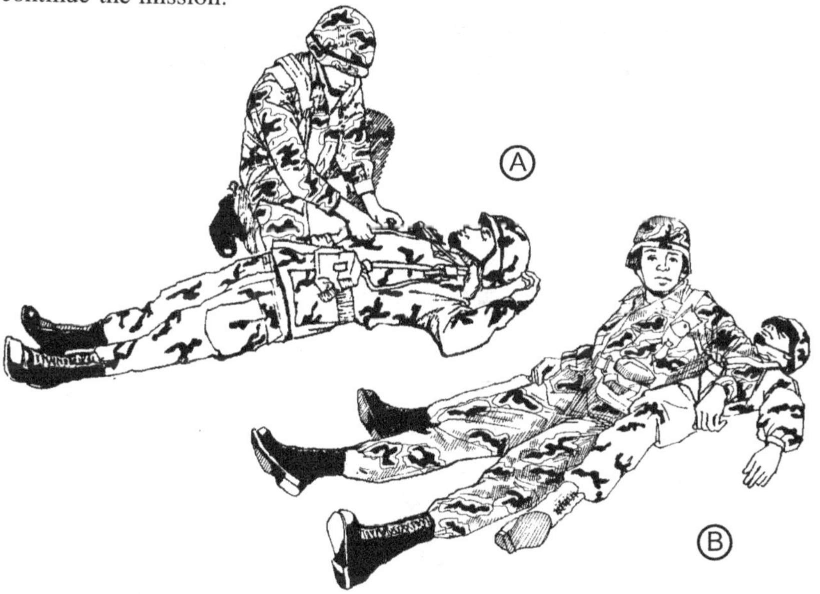

Figure B-15. Load bearing equipment carry using casualty's LBE (Illustrated A—G).

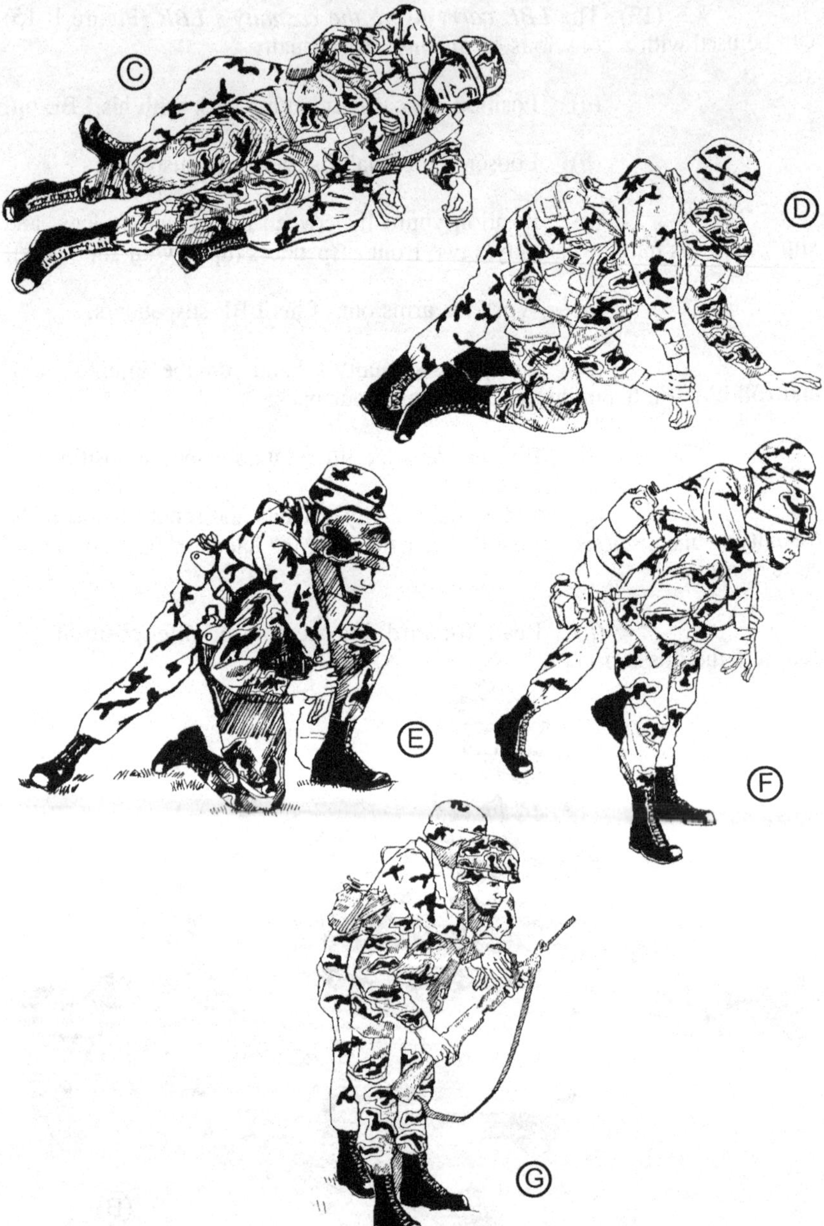

Figure B-15. Load bearing equipment carry using casualty's LBE (Illustrated A—G) (Continued).

b. *Two-man Carries*. These carries should be used whenever possible. They provide more casualty comfort, are less likely to aggravate injuries, and are less tiring for the bearers. Five different two-man carries can be used.

(1) The *two-man support carry* (Figure B-16) can be used in transporting either conscious or unconscious casualties. If the casualty is taller than the bearers, it may be necessary for the bearers to lift the casualty's legs and let them rest on their forearms. The bearers—

(*a*) Help the casualty to his feet and support him with their arms around his waist.

(*b*) Grasp the casualty's wrists and draw his arms around their necks.

Figure B-16. Two-man supporting carry.

(2) The *two-man arms carry* (Figure B-17) is useful in carrying a casualty for a moderate distance (50 to 300 meters) and placing him on a litter. To lessen fatigue, the bearers should carry the casualty high and as close to their chests as possible. In extreme emergencies when there is no time to obtain a spine board, this carry is the safest one for transporting a casualty with a back injury. If possible, two additional bearers should be used to keep the casualty's head and legs in alignment with his body. The bearers—

(*a*) Kneel at one side of the casualty; then they place their arms beneath the casualty's back, waist, hips, and knees.

(*b*) Lift the casualty while rising to their knees.

(*c*) Turn the casualty toward their chests, while rising to a standing position. Carry the casualty high to lessen fatigue.

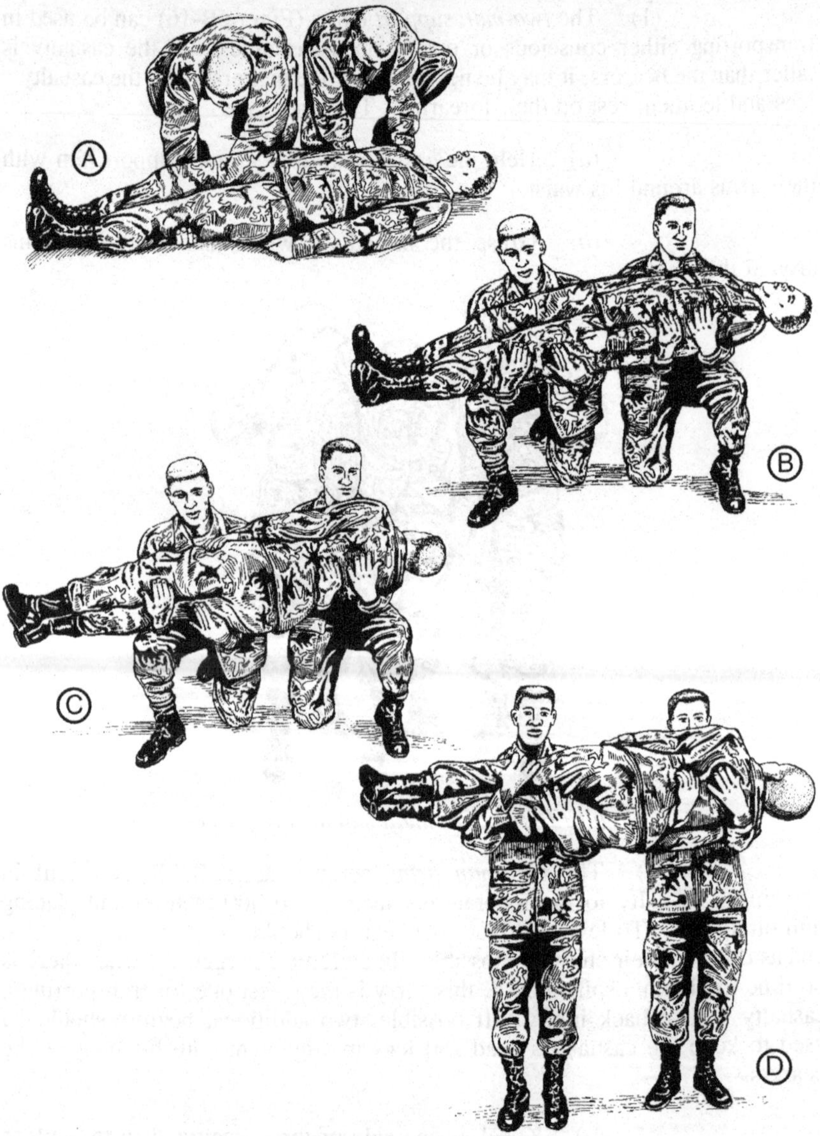

Figure B-17. Two-man arms carry (Illustrated A—D).

(3) The *two-man fore-and aft-carry* (Figure B-18) is a useful *two-man carry* for transporting a casualty for a long distance (over 300 meters). The taller of the two bearers should position himself at the casualty's head. By altering this carry so that both bearers face the casualty, it is useful for placing a casualty on a litter.

(*a*) The shorter bearer spreads the casualty's legs and kneels between them with his back to the casualty. He positions his hands behind the casualty's knees. The other bearer kneels at the casualty's head, slides his hands under the arms, across the chest, and locks his hands together.

(*b*) The two bearers rise together, lifting the casualty.

Figure B-18. Two-man fore-and-aft carry (Illustrated A—B).

(4) Only a conscious casualty can be transported with the *four-hand seat carry* (Figure B-19) because he must help support himself by placing his arms around the bearers' shoulders. This carry is especially useful in transporting a casualty with a head or foot injury for a moderate distance (50 to 300 meters). It is also useful for placing a casualty on a litter.

(*a*) Each bearer grasps one of his wrists and one of the other bearer's wrists, thus forming a packsaddle.

(*b*) The two bearers lower themselves sufficiently for the casualty to sit on the packsaddle; then, they have the casualty place his arms around their shoulders for support. The bearers then rise to an upright position.

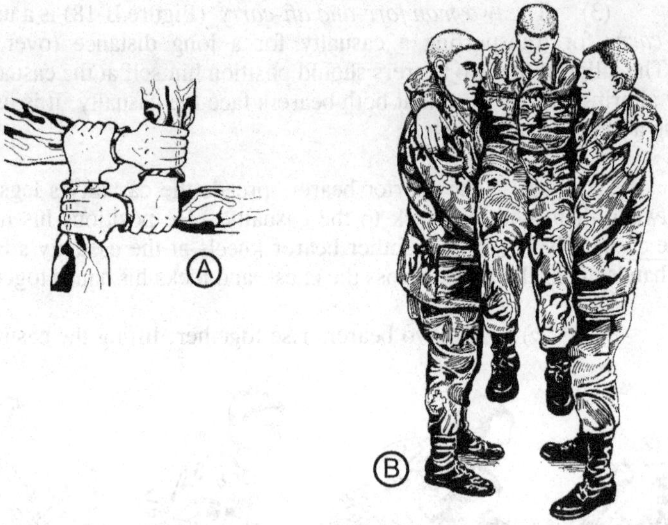

Figure B-19. Four-hand seat carry (Illustrated A—B).

(5) The *two-hand seat carry* (Figure B-20) is used when carrying a casualty for a short distance or for placing him on a litter. With the casualty lying on his back, a bearer kneels on each side of the casualty at his hips. Each bearer passes his arms under the casualty's thighs and back, and grasps the other bearer's wrists. The bearers rise lifting the casualty.

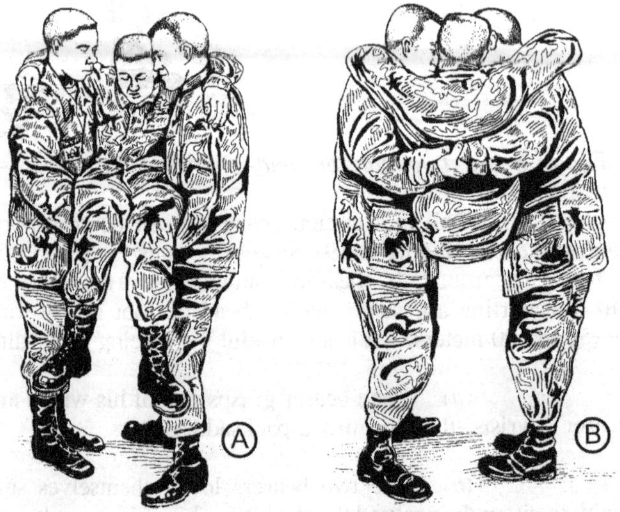

Figure B-20. Two-hand seat carry (Illustrated A—B).

B-9. Improvised Litters

Two men can support or carry a casualty without equipment for only short distances. By using available materials to improvise equipment, the casualty can be transported greater distances by two or more rescuers.

a. There are times when a casualty may have to be moved and a standard litter is not available. The distance may be too great for manual carries or the casualty may have an injury (such as a fractured neck, back, hip, or thigh) that would be aggravated by manual transportation. In these situations, litters can be improvised from materials at hand. Improvised litters must be as well constructed as possible to avoid risk of dropping or further injuring the casualty. Improvised litters are emergency measures and must be replaced by standard litters at the first opportunity.

b. Many different types of litters can be improvised, depending upon the materials available. A satisfactory litter can be made by securing poles inside such items as a blanket, poncho, shelter half, tarpaulin, mattress cover, jacket, shirt, or bed ticks, bags, and sacks (Figure B-18). Poles can be improvised from strong branches, tent supports, skis, lengths of pipe or other objects. If objects for improvising poles are not available, a blanket, poncho, or similar item can be rolled from both sides toward the center so the rolls can be gripped for carrying a patient. Most flat-surface objects of suitable size can be used as litters. Such objects include doors, boards, window shutters, benches, ladders, cots, and chairs. If possible, these objects should be padded for the casualty's comfort.

(1) To improvise a litter using a blanket and poles (Figure B-21), the following steps should be used.

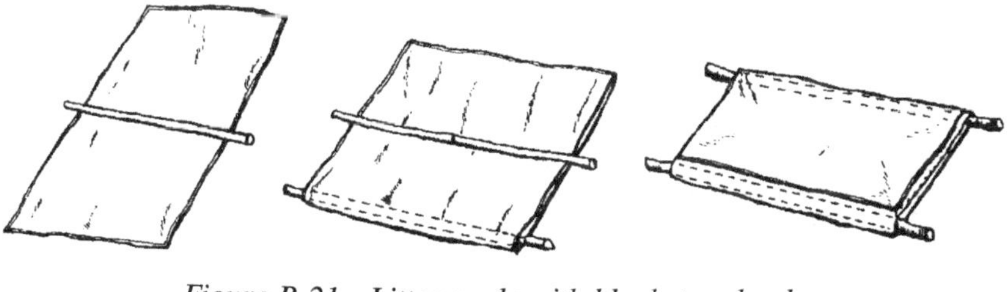

Figure B-21. Litter made with blanket and poles.

(*a*) Open the blanket and lay one pole lengthwise across the center; then fold the blanket over the pole.

(*b*) Place the second pole across the center of the folded blanket.

(*c*) Fold the free edges of the blanket over the second pole and across the first pole.

(2) To improvise a litter using shirts or jackets (Figure B-22), button the shirt or jacket and turn it inside out, leaving the sleeves inside, (more than one shirt or jacket may be required), then pass the pole through the sleeves.

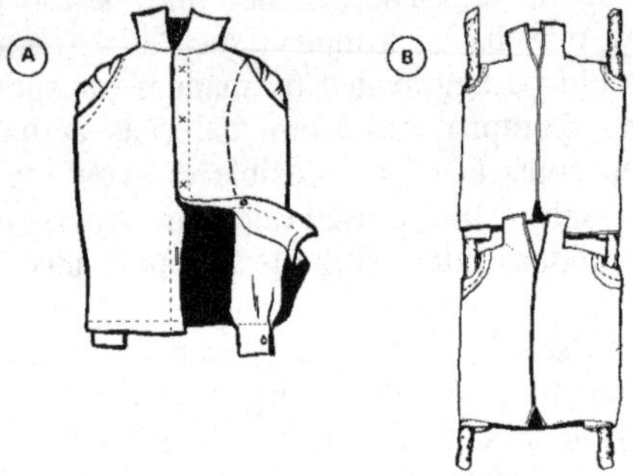

Figure B-22. Litter improvised from jackets and poles (Illustrated A—B).

(3) To improvise a litter from bed sacks and poles (Figure B-23), rip open the corners of bed ticks, bags, or sacks; then pass the poles through them.

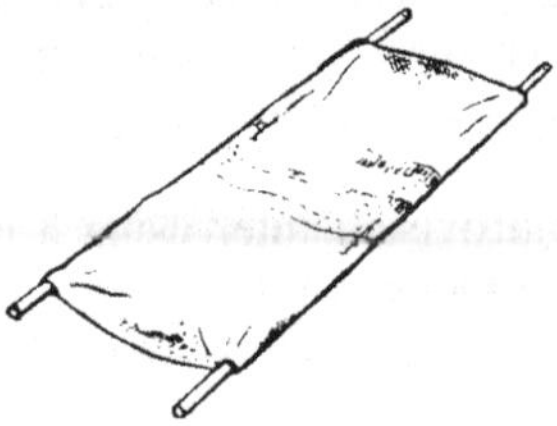

Figure B-23. Litter improvised from bed sacks and poles.

(4) If no poles are available, roll a blanket, shelter half, tarpaulin, or similar item from both sides toward the center (Figure B-24). Grip the rolls to carry the casualty.

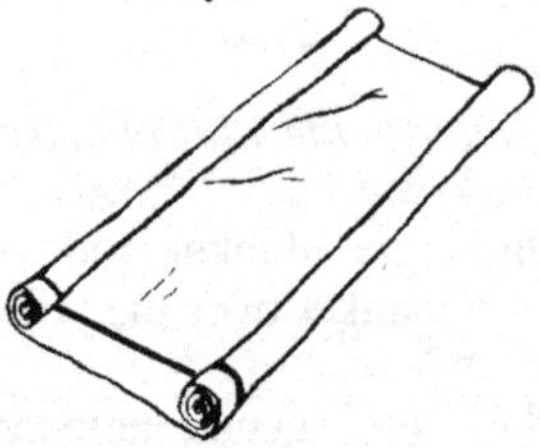

Figure B-24. Rolled blanket used as a litter.

c. Any of the appropriate carries may be used to place a casualty on a litter. These carries are:

- The one-man arms carry (Figure B-6).
- The two-man arms carry (Figure B-17).
- The two-man fore-and-aft carry (Figure B-18).
- The two-hand seat carry (Figure B-20).
- The four-hand seat carry (Figure B-19).

WARNING

Unless there is an immediate life-threatening situation (such as fire, explosion), DO NOT move a casualty with a suspected back or neck injury. Seek medical personnel for guidance on how to transport.

d. Either two or four service members (head/foot) may be used to lift a litter. To lift the litter, follow the procedure below.

(1) Raise the litter at the same time as the other carriers/ bearers.

(2) Keep the casualty as level as possible.

NOTE

Use caution when transporting on a sloping incline/hill.

GLOSSARY

ACRONYMS, ABBREVIATIONS, AND DEFINITIONS

AC hydrogen cyanide
AFMAN Air Force Manual
AOC area of concentration
AR Army regulation
ATM advanced trauma management
ATNAA Antidote Treatment, Nerve Agent, Autoinjector
attn attention

BDO battle dress overgarment
BDU battle dress uniform
BZ anticholinergic drugs

C Celsius
CANA Convulsant Antidote for Nerve Agent
CASEVAC casualty evacuation
cc cubic centimeter
CG phosgene
CHS combat health support
CK cyanogen chloride
Cl chlorine
CLS Combat Lifesaver
CNS central nervous system
CO_2 carbon dioxide
COSR combat and operational stress reactions
CSR combat stress reaction
CTA common table of allowance
CX phosgene oxime

DA Department of the Army
DD Department of Defense
DM diphenylaminochloroarsine (adamsite)
DNBI disease and nonbattle injury
DOD Department of Defense
DP diphosgene
DS direct support

EMT emergency medical treatment

F Fahrenheit
FM field manual

H mustard

HD mustard
HM Hospital Corpsman
HN nitrogen mustard
HSS health service support

IPE individual protective equipment
IV intravenous

JSLIST Joint Services Light Weight Integrated Suit Technology

L lewisite
lasers laser means Light Amplification by Stimulated Emission of Radiation and sources include range finders, weapons/guidance, communication systems, and weapons simulations such as MILES [Multiple Integrated Laser Engagement System].
LBE load bearing equipment
LX lewisite and mustard

MCRP Marine Corps Reference Publication
MILES Multiple Integrated Laser Engagement System
ml milliliter
MOPP mission-oriented protective posture
MOS military occupational specialty
MTF medical treatment facility

NAPP Nerve Agent Pyridostigmine Pretreatment
NATO North Atlantic Treaty Organization
NBC nuclear, biological, and chemical
NCO noncommissioned officer
NTRP Navy Tactical Reference Publication

occlusive dressing air tight transparent dressing used to seal and cover wounds
oz ounce

PAM pamphlet
PS chloropicrin
PTSD post-traumatic stress disorder

QSTAG Quadripartite Standardization Agreement

SOP standing operating procedure
STANAG standardization agreement
STP soldier training publication

2 PAM Cl pralidoxime chloride
TB MED technical bulletin medical
TM technical manual
TSOP tactical standing operating procedure

US United States

WP white phosphorus

REFERENCES

DOCUMENTS NEEDED

These documents must be available to the intended users of this publication.

NATO STANAGs

These agreements are available on request using DD Form 1425 from Standardization Document Order Desk, 700 Robin Avenue, Building 4, Section D, Philadelphia, Pennsylvania 19111-5094.

2122. *Medical Training in First Aid, Basic Hygiene and Emergency Care.* 10 December 1975.
2126. *First Aid Kits and Emergency Medical Care Kits.* 27 September 1983.
2358. *First Aid and Hygiene Training in NBC Operations.* 3 March 1989.
2871. *First Aid Material for Chemical Injuries.* 10 March 1986.

ABCA QSTAGs

These agreements are available on request using DD Form 1425 from Standardization Document Order Desk, 700 Robin Avenue, Building 4, Section D, Philadelphia, Pennsylvania 19111-5094.

535. *Medical Training in First Aid, Basic Hygiene and Emergency Care.* 12 November 1979.

Joint and Multlservice Publications

FM 21-10. *Field Hygiene and Sanitation.* MCRP 4-11.1D. 21 June 2000.

Army Publications

AR 350-41. *Training In Units.* 19 March 1993.
DA PAM 350-59. *Army Correspondence Course Program Catalog.* 26 October 2001.
FM 3-4. *NBC Protection.* FMFM 11-9. 29 May 1992 (Reprinted with basic including Change 1, 28 October 1992; Change 2, 26 February 1996.)
FM 3-5. *NBC Decontamination.* MCWP 3-37.3. 28 July 2000. (Change 1, 31 January 2002.)

FM 3-100. *Chemical Operations Principles and Fundamentals.* MCWP 3-3.7.1. 8 May 1996.
FM 4-02.33 (8-33). *Control of Communicable Diseases Manual* (17th Edition). 3 January 2000.
FM 8-10-6 (4-02.2). *Medical Evacuation in a Theater of Operations—Tactics, Techniques, and Procedures.* 14 April 2000.
FM 4-02.7 (8-10-7). *Health Service Support in a Nuclear, Biological, and Chemical Environment.* 1 October 2002.
FM 8-284 (4-02.284). *Treatment of Biological Warfare Agent Casualties.* NAVMED P-5042; AFMAN (I) 44-156; MCRP 4-11.1C. 17 July 2000. (Change 1, 8 July 2002.)
FM 8-285 (4-02.285). *Treatment of Chemical Agent Casualties and Conventional Military Chemical Injuries.* NAVMED P-5041; AFJMAN 44-149; FMFM 11-11. 22 December 1995.
FM 22-51 (4-02.22). *Leaders' Manual for Combat Stress Control.* 29 September 1994.
CTA 8-100. *Army Medical Department Expendable/Durable Items.* 31 August 1994.
CTA 50-900. *Clothing and Individual Equipment.* 1 September 1994.
STP 21-1-SMCT. *Soldier's Manual of Common Tasks Skill Level I.* 1 October 2001.

Department of Defense Forms

DD Form 1425. *Specifications and Standards Requisition.* March 1986.

READINGS RECOMMENDED

These readings contain relevant supplemental information.

Joint and Multiservice Publications

FM 8-9 (4-02.11). *NATO Handbook on the Medical Aspects of NBC Defensive Operations AMEDP-6 (B), Part I—Nuclear, Part II—Biological, Part III—Chemical.* NAVMED P-5059; AFJMAN 44-151V1V2V3. 1 February 1996.
TB MED 81. *Cold Injury.* NAVMED P-5052-29; AFP 161-11. 30 September 1976.
TB MED 507. *Occupational and Environmental Health Prevention, Treatment and Control of Heat Injury.* NAVMED P-5052-5; AFP 160-1. 25 July 1980.

Army Publications

AR 310-25. *Dictionary of United States Army Terms (Short Title: AD)*. 15 October 1983. (Reprinted with basic including Change 1, 21 May 1986.)

AR 310-50. *Authorized Abbreviations and Brevity Codes*. 15 November 1985.

TM 3-4230-216-10. *Operator's Manual for Decontaminating Kit, Skin: M258A1 (NSN 4230-01-101-3984) and Training Aid, Skin Decontaminating: M58A1 (6910-01-101-1768)*. 17 May 1985. (Change 1, 22 January 1997.)

INDEX

References are to paragraph numbers except where specified otherwise.

PREFACE

1. Scope

This publication describes multi-Service tactics, techniques, and procedures (MTTP) for consideration and use during the tactical employment of nonlethal weapons (NLW) in support of warfighting personnel conducting training and tactical operations. This publication—

a. Provides an overview of NLW and its relationship to deadly force.

b. Provides NLW system description.

c. Describes the capability requirements of NLW.

d. Discusses fundamental concepts and training requirements involved with NLW.

e. Discusses the tactical employment considerations of NLW.

f. Discusses lessons learned from previous NLW use.

2. Purpose

a. This publication provides a single-source, consolidated reference on the tactical employment of NLW and supporting systems.

b. This publication is not intended to restrict the authority of the commander from organizing the force and executing the mission in a manner deemed appropriate.

3. Application

a. This publication provides commanders and their staffs unclassified guidance for NLW employment and planning. Commanders and staffs can use this publication to aid in the tactical employment of NLW during exercises and contingencies.

b. The United States (U.S.) Army, Marine Corps, Navy, Air Force, and Coast Guard approved this multi-Service publication for use.

4. Implementation Plan

Participating service command offices of primary responsibility will review this publication, validate the information and reference, and incorporate it in service manuals, regulations, and curricula follows.

Army. The Army will incorporate the procedures in this publication in U.S. Army training and doctrinal publications as directed by the Commander, U.S. Army Training and Doctrine Command (TRADOC). Distribution of this publication is in accordance with (IAW) initial distribution number (IDN) 115757.

Marine Corps. The Marine Corps will incorporate the procedures in this publication in U.S. Marine Corps doctrinal and training publications as directed by the Commanding General, U.S. Marine Corps Combat Development Command (MCCDC). Distribution of this publication is IAW Marine Corps Doctrinal Publication System (MCPDS).

Navy. The Navy will incorporate the procedures in this publication in U.S. Navy doctrinal and training publications as directed by the Commander, Navy Warfare Development Command (NWDC). Distribution of this publication is IAW Military Standard Requisition-

MARINE CORPS PCN: 143 000057 00
Air Force Distribution: F

ing and Issue Procedures (MILSTRIP) Desk Guide and Navy Standing Operating Procedure (NAVSOP) Pub 409.

Air Force. The Air Force will validate and incorporate appropriate procedures IAW applicable governing procedures. Distribution is IAW Air Force Instruction (AFI) 33-360.

Coast Guard. The Coast Guard will incorporate the procedures in this publication in U.S. Coast Guard doctrinal and training publications as directed by the Assistant Commandant for Operations, Director of Operations Policy, U.S. Coast Guard (USCG). Distribution of this publication is IAW Coast Guard standing operating procedures.

5. User Information

a. TRADOC, MCCDC, NWDC, Air Force Doctrine Center (AFDC), USCG, and the Air Land Sea Application (ALSA) Center developed this publication with the joint participation of the approving service commands. ALSA will review and update this publication as necessary.

b. This publication reflects current joint and service doctrine, command and control organizations, facilities, personnel, responsibilities, and procedures regarding NLW employment. Changes in service protocol, appropriately reflected in joint and service publications, will likewise be incorporated in revisions of this document.

c. We encourage recommended changes for improving this publication. Key comments to the specific page and paragraph and provide rationale for each recommendation directly to—

*FM 3-22.40 (FM 90-40)
*MCWP 3-15.8
*NTTP 3-07.3.2
AFTTP(I) 3-2.45
*USCG PUB 3-07.31

FM 3-22.40 (FM 90-40)	U.S. Army Training and Doctrine Command Fort Monroe, Virginia
MCWP 3-15.8	U.S. Marine Corps Combat Development Command Quantico, Virginia
NTTP 3-07.3.2	U.S. Navy Warfare Development Command Newport, Rhode Island
AFTTP(I) 3-2.45	Headquarters, Air Force Doctrine Center Maxwell Air Force Base, Alabama
USCG Pub 3-07.31	U.S. Coast Guard Washington, D.C.

15 January 2003

NLW

Tactical Employment of Nonlethal Weapons

TABLE OF CONTENTS

Page

*This publication supersedes FM 90-40, MCRP 3-15.8, NWP 3-07.31, USCG Pub 3-07.31, dated 6 October 1998.

FIGURES

Tables

EXECUTIVE SUMMARY

Tactical Employment of Nonlethal Weapons (NLW)

Today's world events mandate a need to project nonlethal force across all levels of war to enable our warfighters and leaders to effectively deal with a host of traditional as well as nontraditional threats. Now more than ever, the minimal level of tolerance for collateral damage and loss of human life, coupled with the tendency for the typical adversary to exploit the rules of engagement (ROE) to his benefit, necessitates an effective and flexible application of force through nonlethal weapons. Much like a rheostat switch in which power can be dialed up or down as desired, NLW provide tools to allow a commander to employ sufficient force to accomplish an objective without requiring the destruction of an enemy or the habitat. The intent of employing NLW is not to add another step in the progression of escalation with an adversary, but to add another tool to use anywhere along that continuum. This publication provides initial guidance for employing NLW in a tactical environment.

Overview of NLW

Chapter I describes the background, definition, and the concept description of NLW. This chapter discusses the development and employment of NLW, and explains the relationship of NLW to deadly force capabilities. The final portion of the chapter reviews policy per Office of Secretary of Defense instructions on employing NLW.

Purpose, System Description, and Organization

Chapter II describes the purpose of NLW and gives a general description of currently available commercial or government off-the-shelf equipment and munitions. The chapter also addresses organization structure, to include system location/distribution, personnel requirements, and mission requirements/tactical structure.

New Capability Requirements

Chapter III describes the applicability of NLW across the range of military operations. Additionally, this chapter discusses the core capabilities as stated in the Joint Mission Area Analysis (JMAA) as described in the Department of Defense Joint Concept for NLW.

Training With the NLW Capability Set

Chapter IV describes training requirements, topics, methods, locations, and a program of instruction from the Interservice Nonlethal Individual Weapons Instructor Course currently available for NLW options. This chapter also addresses additional training available from civilian law enforcement agencies.

Employment of NLW Capabilities

Chapter V establishes the fundamental understanding of the force continuum needed to employ NLW. It outlines employment prerequisites, mission-planning factors, and personnel requirements needed to employ NLW options. It further delineates NLW tasks according to the JMAA and provides considerations for preparation for unit tactical operations. This chapter also discusses employment objectives for NLW and ROE development. This chapter ends with multiple examples of service-specific crowd control formations used by units equipped with NLW.

Lessons Learned

Chapter VI focuses on NLW lessons learned from recent military operations. The chapter discusses how nonlethal technology has changed the way forces are looking at their adversaries in using NLW. It also covers the use of NLW along the force continuum in conjunction with ROE and the media.

PROGRAM PARTICIPANTS

The following commands and agencies participated in developing this publication:

Joint

U.S. Joint Forces Command, Suffolk, VA
U.S. European Command, J3/J5, APO AE
U.S. Pacific Command, J3832, Camp H. M. Smith, HI
U.S. Central Command, J3, MacDill AFB, FL
U.S. Southern Command, SCJ5-PS, Miami, FL
U.S. Space Command, SPJ5X, Peterson AFB, CO
U.S. Special Operations Command, HQ USSOCOM, SOOP-PTE, MacDill AFB, FL
Joint Nonlethal Weapons Directorate, Quantico, VA
Interservice Nonlethal Individual Weapons Instructor Course, Fort Leonard Wood, MO

Army

Headquarters, Department of the Army, ODCSOPS (DAMO-FDT), Pentagon, Washington, D.C.
U.S. Army Training and Doctrine Command, DCSDCS (ATDO-A), Fort Monroe, VA
U.S. Army Training and Doctrine Command, DCSDEV (ATCD-SN), Fort Monroe, VA
U.S. Army Infantry Center, DCD, Fort Benning, GA
U.S. Army Military Police School, ATTN: DOTD, MANSCEN (ATZT-DT-WM-M), Fort Leonard Wood, MO
USAJFKSWCS, DOTD, Fort Bragg, NC
National Ground Intelligence Center, Charlottesville, VA

Marine Corps

Marine Corps Combat Development Command, Quantico, VA
MARCORSYSCOM, Quantico, VA
USMC Warfighting Lab, Quantico, VA
MCSFT Company, MCSF Battalion, Chesapeake, VA
2d FAST MCSF Battalion, Yorktown, VA
1st MARDIV Military Police Company, HQ Battalion, Camp Pendleton, CA
Marine Corps Detachment Military Police and Chemical School, Fort Leonard Wood, MO
Special Operations Training Group, I Marine Expeditionary Force, Camp Pendleton, CA
Special Operations Training Group, II Marine Expeditionary Force, Camp Lejeune, NC

Navy

Commander, Navy Warfare Development Command (NWDC), Newport, RI
USN Surface Warfare Center, Crane, IN
USN Surface Warfare Center, Dahlgren, VA

Air Force

AFDC/DJ, Langley AFB, VA
Headquarters, ACC/SFOT, Langley AFB, VA
Air Force Security Forces Center/SFOR, Lackland, AFB, TX
Headquarters, USAF/XOF, Washington, D.C.

Coast Guard

Commandant, U.S. Coast Guard, (G-OPL-2), Washington, D.C.

Chapter I

Overview of Nonlethal Weapons

1. Background

a. In today's operational environment, United States (U.S.) forces regularly perform missions that were almost inconceivable a few decades ago. In this new environment, lethal firepower or the threat of its use may no longer be the default solution to all crises or problems. Senior leaders face a new level of public and media sensitivity and scrutiny concerning the proper role of the military as an instrument of national power. Field commanders must understand these sensitivities and attempt to achieve an appropriate military force. Junior leaders must apply the resulting decisions wisely, often in changing situations filled with uncertainty and danger.

b. Throughout history, changes in culture and technology have influenced the character of military force and the manner in which it is employed. These changes are an attempt to maximize military force in the force continuum. In the complex and changing political and social domain of the modern world, nonlethal capabilities may offer the opportunity to increase or maximize the military as an instrument of national power.

c. Nonlethal weapons (NLW) must now be considered an integral part of joint operational capabilities. Joint forces anticipate that NLW technology will advance and provide additional options for force commanders.

2. Definition

a. Department of Defense Directive (DODD) 3000.3, Policy for Nonlethal Weapons, 9 July 1996, defines NLW as "weapons that are explicitly designed and primarily employed so as to incapacitate personnel or materiel while minimizing fatalities, permanent injury to personnel, and undesired damage to property and the environment." Furthermore, "unlike conventional lethal weapons that destroy their targets principally through blast, penetration, and fragmentation, nonlethal weapons employ means other than gross physical destruction to prevent the target from functioning. Nonlethal weapons are intended to have relatively reversible effects on personnel and materiel."

b. The term "nonlethal" does not mean zero mortality or nonpermanent damage; these are goals and not guarantees of these weapons. NLW can add flexibility to combat operations and enhance force protection by providing an environment in which friendly troops can engage threatening targets with reduced risk of noncombatant casualties and collateral damage.

3. Concept Description

Nonlethal capabilities within the context of the service capability sets (see Appendix E) may provide armed forces additional offensive and defensive tools to assist in the incapacitation and destruction of an enemy, capture of important intelligence sources, crowd control, civil disturbances, enforcing mandates, and force protection in accordance with the specified rules of engagement (ROE) for a specific operation. These operations include small-scale contingencies (SSC), military operations other than war (MOOTW), stability and support operations (SASO), and civil disturbances. The U.S. Army's Task Force (TF) FALCON in Kosovo and U.S. armed forces located at Camp X-RAY in Guantanamo Bay, Cuba are superb examples of the use of NLW to prevent escalation within an operation.

4. Background on Development and Employment

a. Military forces have long used nonlethal force to influence behavior of people and nations, defeat adversaries with minimum use of lethal capabilities, and weaken adversaries to defeat them with conventional arms. Examples of nonlethal means include a show of force, physical obstacles, riot control agents (RCA), noise to create or enhance psychological effects, smoke and obscurants to mask operations or defeat homing and guidance mechanisms, and using light to disorient combatants. These nonlethal means will remain relevant in future operations and are a part of an evolutionary process of weapons development. Joint operations in Somalia provided increased awareness of the need to incorporate nonlethal options into the fighting capabilities. Therefore, the Department of Defense (DOD) established the Joint Nonlethal Weapons Program in 1996 to research, develop, and acquire these capabilities.

b. Increased interaction between U.S. forces and civilian populations has become a feature of the contemporary operational landscape that is likely to remain the case in the future. Three factors account for this forecast:

(1) Worldwide patterns of population growth and migration have resulted in an expansion of urban culture within established industrialized nations, as well as many pre-industrial and emerging third world societies (60 percent of the world's population live in built-up areas). The prevalence of urbanization in many crisis-prone regions of the world creates the potential for large, vulnerable groups of noncombatants to be entrapped in volatile confrontations involving the deployment of military forces in military operations in urban terrain (MOUT).

(2) U.S. forces increasingly operate in the challenging environment known as MOOTW; a category that includes such missions as humanitarian assistance, disaster relief, noncombatant evacuations, and the various types of peace operations. These operations commonly involve close and continual interaction between U.S. forces and noncombatant civilians. Some MOOTW scenarios include the presence of paramilitary forces, armed factions, or rogue elements that present a constant, but uncertain threat to joint forces. In these situations, the mission of military forces includes preventive tasks. In other words, U.S. forces enhance mission accomplishment by preventing individuals or factions from carrying out specified undesirable activities, such as rioting, looting, attacking, harassing, or threatening. Often, the adversary blends in with the local population of innocent citizens. At other times, sectors of the local population may rise against U.S. forces and become active participants in acts of violence. Factional alignments, the level of violence, and the threat to mission accomplishment, may change frequently with little or no warning. Under such circumstances, the identity of adversaries is uncertain and the use of deadly force for purposes other than self-defense may be constrained by ROE or the judgment of the on-scene commander.

(3) From a defense-in-depth or situational understanding perspective, the force protection, homeland security, and the antiterrorism missions, require that forces assess the use of NLW as a response to potential asymmetric attacks. Asymmetric threats run the spectrum of adversarial use of innovative or nontraditional strategies, tactics, or technologies, to exploit vulnerabilities and strike a decisive blow while avoiding the strengths of an opponent. The objectives of these strikes are intended to undermine national will. Terrorist attacks such as those directed against the USS Cole, World Trade Center, and Pentagon, highlight the severity of the asymmetric threat and its resultant effects.

5. Relationship of NLW to Deadly Force

a. NLW provide "the lesser means" that could be considered prior to employing deadly force. The services definition of deadly force is as follows:

(1) U.S. Army – "Force that a person uses causing, or that a person knows, or should know, will create a substantial risk of causing death or serious bodily harm." (Army Regulation (AR) 190–14, Military Police Carrying of Firearms and Use of Force for Law Enforcement and Security Duties, 12 March 1993).

(2) USMC – "That force which a person uses with the purpose of causing death or serious bodily harm or which a reasonable and prudent person would consider likely to create substantial risk of causing death or serious bodily harm. It will be used as a last resort when all lesser means have failed." Marine Corps Order (MCO) 5500.6F, Arming of Security and Law Enforcement Personnel and the Use of Force, 20 July 1995.

(3) USN – "Force that a person uses causing, or that a person knows or should know would create a substantial risk of causing death or serious bodily harm." Secretary of The Navy Instruction (SECNAVINST) 5500.29B, Use of Deadly Force and the Carrying of Firearms by Personnel of the Department of the Navy in Conjunction with Law Enforcement, Security Duties and Personal Protection, 29 September 1992.

(4) USAF – "Force that causes, or that you know has a substantial risk of causing death or serious bodily harm." Air Force Instruction (AFI) 31-207, Arming and Use of Force by Air Force Personnel, 1 September 1999.

(5) USCG – "Deadly force is any force that is likely to cause death or other serious physical injury." (COMDTINST M16247.1B, Maritime Law Enforcement Manual, 9 August 2001.)

b. The commitment of military power to resolve crises has traditionally involved the use of deadly force or the implicit or explicit threat of the use of deadly force. Military units are trained, organized, and equipped for this purpose. A force armed only with traditional military weapons normally has two options for enforcing compliance: threating deadly force and applying deadly force. This limitation creates a critical vulnerability that belligerents may quickly discern and use to their advantage.

c. Nonlethal capabilities provide a wider range of options that augment, but do not replace traditional means of deadly force. The option to resort to deadly force must always remain available when the commander believes it is appropriate to the mission. DODD 3000.3, states "the availability of NLW shall not limit a commander's inherent authority and obligation to use all necessary means available and to take all appropriate action in self-defense." The existence of nonlethal capabilities does not represent the potential for "nonlethal war," and unrealistic expectations to that effect must be vigorously avoided. Noncombatant casualties, to include serious injuries and fatalities, are often an unavoidable outcome when employing military power, regardless of NLW availability. Augmenting deadly force with NLW should be fundamental to planning and executing any operation where the commander is contemplating deadly force.

d. The commander must clearly articulate the roles of NLW within the ROE. NLW is an additional means of employing force throughout the force continuum for the particular purpose of reducing the probability of death or serious injury to noncombatants or belligerents. Using deadly force must always remain an inherent right of individuals in instances when they, their fellow servicemen, or personnel in their charge are threatened with death or serious bodily harm. NLW add flexibility to combat operations and enhance force protection by providing the field commander with options to engage threatening targets with limited risk of noncombatant casualties and collateral damage.

e. Commanders and public affairs officers (PAO) must prepare personnel to address media questions and concerns regarding the role of NLW. Operational experience indicates that new capabilities give rise to significant media interest. Personnel participating in interviews or briefings must be prepared to address the role of NLW. It must be clear that the presence of NLW in no way indicates abandoning the option to employ deadly force in appropriate circumstances.

6. Advantages of Employing NLW

NLW provide commanders the flexibility to influence the situation favorably with increased safety to friendly forces and reduced risk of both noncombatant fatalities and collateral damage. Some advantages follow:

a. NLW can be more humane and consistent with the political and social implications implicit in humanitarian and peacekeeping missions.

b. The force that properly employs nonlethal options can gain advantages over those who rely on lethal options alone. Because the degree of provocation required to employ NLW is substantially less than lethal options, the force can take a more proactive position that may diminish the likelihood of having a situation escalate to a point where deadly force is required to resolve a conflict.

c. NLW options can be less likely to provoke others. However, commanders should be cautioned using NLW does not preclude an undesired result.

d. Demonstrated restraint can greatly diminish feelings of anger and remorse when deadly force is required after nonlethal options fail.

e. Using NLW can facilitate post incident stabilization by reducing populace alienation and collateral damage.

f. NLW can reduce the possibility of injury to friendly forces compared to forces without NLW.

7. Policy Per Office of the Secretary of Defense Instructions on Employment of NLW

a. DODD 3000.3 states, "nonlethal weapons, doctrine, and concepts of operation shall be designed to reinforce deterrence and expand the range of options available to commanders. Nonlethal weapons should enhance the capability of U.S. Forces to accomplish the following objectives:

(1) Discourage, delay, or prevent hostile actions.

(2) Limit escalation.

(3) Take military action in situations where use of lethal force is not the preferred option.

(4) Better protect our forces.

(5) Temporarily disable equipment, facilities, and personnel.

(6) Nonlethal weapons should also be designed to help decrease the post-conflict costs of reconstruction."

b. DODD 3000.3 continues by stating, "The availability of nonlethal weapons shall not limit a commander's inherent authority and obligation to use all necessary means available and to take all appropriate action in self-defense. Neither the presence nor the potential effect of nonlethal weapons shall constitute an obligation for their employment or a higher standard for employment of force than provided for by applicable law. In all cases, the United States retains the option for immediate use of lethal weapons, when appropriate, consistent with international law."

c. DODD 3000.3 further states, "Nonlethal weapons shall not be required to have a zero probability of producing fatalities or permanent injuries. However, while complete avoidance of these effects is not guaranteed or expected, when properly employed, nonlethal weapons should significantly reduce them as compared with physically destroying the same target." Nonlethal weapons may be used in conjunction with lethal weapon systems to enhance the latter's effectiveness and efficiency in military operations. This shall apply across the range of military operations to include those situations where overwhelming force is employed."

d. According to paragraph 3.21 of the Interium Guidance issued by the Deputy Secretary of Defense, 30 October 2002, all developmental weapon systems must be submitted for legal review and must receive favorable findings before they can be incorporated into the force. This review includes examination for consistency with applicable international and domestic laws, including the laws of war, various arms control treaties, and protocols. U.S. forces can be assured that any NLW they are issued have passed this critical test. Medical and policy reviews should also be considered prior to fielding any new weapon systems. Commanders and operators should consult their specific ROE for guidance.

e. Some capabilities, although designed to minimize fatalities and serious injuries, may have effects that could actually discourage their use. Therefore, the basic decision to employ military force in defense of national interests is usually a matter of intense public concern. The manner in which that force is exercised is subjected to the same scrutiny.

f. The presence of NLW on the battlefield does not mean they must be used. It also does not mean they must be used prior to the use of lethal force. In all cases, the U.S. retains the option for immediate use of lethal weapons when the on-scene commander deems it appropriate. Additionally, the immediate use of lethal weapons must be consistent with U.S. and international law and the ROE.

Chapter II

PURPOSE, SYSTEM DESCRIPTION, AND ORGANIZATION

1. Purpose

The purpose of NLW is to enhance the ability of military commanders to conduct operations across the spectrum of conflict. Over the past several years, military operations have gravitated towards the lower end of this spectrum and have involved U.S. forces in various missions. These missions include SSC, SASO, MOOTW, and civil disturbances where the ROE are inherently restrictive. NLW are designed to address the special requirements of low-intensity environments. They provide tools that enable the individual, team, or unit, to anticipate and/or respond to provocation with more appropriate means than through the use of deadly force. Essentially, NLW provide U.S. forces with a near-term capability for effectively controlling the nontraditional battlefield, within the constraints levied by ROE, by reducing the risk to friendly forces, mitigating casualties, and minimizing collateral damage.

2. System Description

a. NLW Capability Set. The services have, and are continuing to procure, versatile packages of NLW comprised of commercial off-the-shelf (COTS) and government off-the-shelf (GOTS) equipment and munitions. The government is also exploring advanced technologies for future procurements. In addition to addressing contingency requirements, the services' NLW capability sets provide limited sustainment training ammunition and equipment.

b. NLW Capability Set—Component Categories. Each service determines the contents of its own capability set. Examples of the services current configurations are shown in Appendix E. The NLW capability set components can be divided into four distinct categories: personnel protectors, personnel effectors, mission enhancers, and training devices. The following examples are not all-inclusive.

(1) Personnel Protectors. Personnel protectors include items such as face shields, riot shields, and shin guards that protect the individual from blunt trauma injuries inflicted by thrown objects, clubs, etc.

(2) Personnel Effectors. Personnel effectors include items such as riot batons, malodorants, stingball grenades, pepper sprays, and other kinetic rounds (e.g. sponge grenades) designed to discourage, disorient, or incapacitate individuals or groups.

(3) Mission Enhancers. Mission enhancers include items such as bullhorns, spotlights, caltrops, isotropic radiators (stun/flash), and barriers/entanglers. These items are designed to facilitate target identification and crowd control, and to limit personnel and vehicular movement.

(4) Training Devices. Training devices include items such as training suits, training rounds, training batons, and inert pepper sprays. They are designed to facilitate realistic hands-on scenario training in preparation for operations.

c. System Relationships and Inter/Intraoperability. The NLW capability set is designed to provide an operating force (ranging from squad through battalion-sized units [see Appendix E]) with tactically deployable nonlethal equipment. Fielding of this capability set enhances the operation of systems and equipment already fielded and planned within the joint community. The NLW capability set components are compatible with military equipment that is already fielded or planned for future fielding. The components of the NLW capability set are supported within current military logistic capabilities.

d. Augmentation of Organic Equipment. The operational intent of the capability set is to augment an operating force with the required nonlethal equipment to enhance its force projection capability. The NLW capability set is additional equipment that must be incorporated into lift and load planning.

3. Organization

Organization/organizational structure is not anticipated to change as a result of NLW capability sets being added to the inventory.

a. System Location/Distribution. The United States Army (USA), United States Marine Corps (USMC), and United States Navy (USN) have procured and fielded capability sets to the operating forces, security forces, bases, and stations. All services have current plans to field additional capability sets as required. At the time of this printing, the United States Air Force (USAF) has identified its capability sets to field in the near future. The United States Coast Guard (USCG) also intends to field NLW.

b. NLW Capability Set Composition. Appendix E describes the current and proposed capability sets.

c. Personnel Requirements. To date, no change or increase in military specialty codes have been determined in order to field, operate, and maintain the NLW capability set. Each service determines its own training personnel requirements.

d. Mission Requirements/Tactical Structure.

(1) The mission requirements and tactical structure pertinent to the employment of the NLW will vary depending on the type of contingency and subsequent task organization of the employing unit. Mission requirements, derived from commander's guidance, should be satisfied through the appropriate combination of lethal and nonlethal capabilities.

(2) NLW provide commanders with a means of shaping their area of operations (AO) short of employing deadly force. However, it is imperative that U.S. forces retain the means and ability to employ lethal force in the event that nonlethal force fails to achieve desired outcomes. For example, NLW employment may provoke a lethal response from the adversary. Without the imminent threat of lethal force, nonlethal munitions will prove to be less effective and possibly even counterproductive. More importantly, denying U.S. forces a lethal capability would place them at unacceptable risk. If nonlethal force is employed without the threat of lethal force, the belligerent may opt to exploit a perceived vulnerability. In this case, the use of NLW becomes counterproductive, because it serves only to aggravate the threat and causes the situation to deteriorate.

Chapter III

NLW CAPABILITY REQUIREMENTS

1. Applicability Across the Range of Military Operations

a. U.S. forces conduct military operations across a large spectrum. This spectrum spans from humanitarian assistance and peacekeeping to strategic nuclear exchange. Since the end of the Cold War, the U.S. military has been increasingly involved with operations at the lower end of the spectrum. These operations focus on deterring war, resolving conflict, supporting civil authorities and promoting peace. NLW may be useful across many categories of these operations. The range of military operations includes various levels of combat "intensity," and NLW provide one more force option to deal with the varying intensity. MOOTW may be found across the range of military operations and often include situations where noncombatants are mingled with the adversaries or involved in acts of violence, such as rioting or looting. In such circumstances, NLW provide commanders the flexibility to influence the situation favorably with reduced risk of noncombatant fatalities and collateral damage. NLW also provide a means by which to ascertain the intent of individuals/belligerents. Some characteristics of operations that present the opportunity to use NLW are—

(1) Close contact between friendly forces and noncombatant civilians.

(2) The presence of paramilitary or irregular forces.

(3) Armed civilian gangs, unlawful combatants, or militia.

b. These characteristics present a real but ill-defined threat. Under such circumstances, the commander's judgment or ROE may restrict using deadly force for other than self-defense. NLW may provide U.S. forces flexibility to resolve situations where the use of deadly force is restricted. These weapons provide flexibility by—

(1) Allowing U.S. forces to apply measured military force.

(2) Reducing risk of serious noncombatant causalities.

(3) Accomplishing the mission.

(4) Providing force protection.

(5) Retaining the initiative.

(6) Allowing U.S. forces to remain relevant and capable in increasingly complex and chaotic environments.

c. The need to reduce the risks of serious injury to personnel is not limited to crowd control scenarios. In MOUT, some of the local civilian populace may remain in an urban area in the midst of battle. The traditional solution to such challenges has been restricting ROE. Nonlethal capabilities may provide commanders the flexibility to adapt a fluid approach to MOUT, which allows their subordinates the freedom of action to employ appropriate levels of measured military force to accomplish their mission.

d. Another example of NLW applications in MOOTW is in peace enforcement. In these operations, NLW are used in an area denial role to separate belligerents without resorting to deadly force. They are also used to support humanitarian relief organizations in protecting food and medical distribution sites. The presence of NLW will not limit a commander's inherent authority and obligation to use all necessary means available and to take appropriate action in self-defense.

2. Core Capabilities

Core capabilities are those fundamental competencies that enable U.S. forces to achieve desired operational objectives. NLW provide a flexible means to protect friendly forces, high-value assets, and influence the actions of potential adversaries and/or noncombatants. These goals can be achieved without resorting to lethal force and in a manner that may minimize collateral damage. The core capabilities associated with nonlethal effects fall into three major categories as stated within the Joint Mission Area Analysis (JMAA) for Nonlethal Weapons (approved December 2000 by the Joint Requirements Oversight Council): counterpersonnel, countermateriel, and countercapability.

a. Counterpersonnel Capabilities. The DOD Joint Concept for Nonlethal Weapons concept paper, 5 January 1998, describes required nonlethal counterpersonnel capabilities for U.S. forces as follows:

(1) "The application of military force to accomplish a mission with reduced risk of fatalities or serious casualties among noncombatants—or even, in some instances, among enemy forces." There are several specific nonlethal counterpersonnel capabilities to be considered. These include, but are not limited to, the means to influence the behavior and activities of a potentially hostile crowd and the capability to bring a mob engaged in a riot under control. While there are many similarities in these two capabilities, each involves unique challenges, which may require radically different solutions.

(2) "Capability to incapacitate personnel." For the purposes of this document, "incapacitation" is achieved when the weapon's effects render personnel unable to physically or mentally (real or perceived) act in a hostile or threatening manner. In keeping with the guiding principles for NLW, this incapacitation should be reversible through the passage of time. The focus will be on capabilities that affect both individuals and groups of people.

(3) "Capability to deny personnel access to an area (land, sea, or air)." NLW can provide new possibilities for barrier planning in any type of military operation, including systems that produce physical or mental discomfort to those who enter the denied area.

(4) "Capability to clear facilities and structures of personnel." The Joint Concept for NLW states this capability "will facilitate MOUT by reducing the risks of noncombatant casualties and collateral damage while simultaneously minimizing the advantages accruing to an enemy defending a built up area." This capability is applicable to all clearing missions that may be conducted (e.g., Maritime Interception Operations [MIO]) and could assist in providing enhanced force protection.

(5) "Capability to seize personnel, which could include some combination of technologies inherent in other core capabilities, for example, incapacitation and entangling devices (which may aid in seizing personnel)." This capability is intended to augment lethal means used to capture specified individuals, such as enemy combatants or persons who are inciting a mob or crowd to violence.

b. Countermateriel Capabilities. The Joint Concept for Nonlethal Weapons describes countermateriel capabilities as follows:

(1) Rendering equipment and facilities unusable without complete destruction by attacking only weapons of war and supporting infrastructure while minimizing risk to noncombatants. This could enhance operations by defusing potentially volatile situations where conventional military means might otherwise prove counterproductive.

(2) Disabling or denying the use of vehicles, vessels, and aircraft entry into targeted areas or access within an AO. This capability may include physical barriers that impede or stop movement. Alternative systems may alter combustion properties of fuels

and viscosity of lubricants, embrittle or decay materials, attack tires, gaskets, or hoses, or act as adhesives to fuse metal parts. The requirement also includes the capability to facilitate boarding and clearing operations.

c. Counter Capability Capabilities. For U.S. forces counter capability capabilities are described within the JMAA for NLW as follows:

(1) "Disable or neutralize facilities and systems." This includes a nonlethal capability for disabling/neutralizing electrical generating facilities, command, control, communications, computers, intelligence, surveillance, and reconnaissance (C4ISR) systems, Integrated Air Defense Systems (IADS), weapons systems, optical sensors, electrical sensors, and navigation capabilities with such controlled effects as to allow for selective, precise engagement. Other needs include deceiving reconnaissance with illusions of barriers, obstacles, and/or forces, and concealing/covering friendly activities or movements, particularly within complex urban terrain.

(2) "Deny the use of weapons of mass destruction (WMD)." Mission needs include rendering a WMD inoperative; containing the potential release of deadly agents/contaminants; and preventing or neutralizing the production, storage, deployment (transport), employment, and delivery of WMD. Nonlethal capabilities in combating WMD should be designed to permit engagement within a populated or sensitive terrain, prior to the initiation of armed conflict.

3. Limitations

The limitations of NLW are effectiveness, perception, and legality.

a. The effectiveness of NLW is dependent on factors such as motivation, age, environmental and human factors, range of engagement, and numerous others. NLW can be lethal if improperly applied. Furthermore, countermeasures for thwarting virtually all nonlethal options are usually apparent, quickly learned, and readily available. Because they are not intended to kill, nonlethal options teach adversaries what to avoid in the future. Small unit commanders must prepare to keep one step ahead of belligerents.

b. The perceptual limitation is that personnel may misunderstand the appropriate applications of NLW across the range of military operations. The incorrect perception that NLW will allow wars and MOOTW to be prosecuted without casualties may lead to conflicting expectations between political and military leaders. These conflicts can result in U.S. forces being vulnerable to misplaced or inconsistent ROE. All leaders, political and military, involved in planning and executing military missions must understand there are no "nonlethal operations."

c. Some proposed NLW might be forbidden by law or policy. Accordingly, it is essential that all NLW developments be evaluated by appropriate authorities to ensure they comply with the law of war, U.S. law, and U.S. treaty obligations. Use of NLW containing chemical agents, for instance, must be evaluated in the context of the Chemical Weapons Convention. Also, using RCA in an armed conflict requires Presidential approval.

Chapter IV

TRAINING

1. Fundamental Concepts and Training Requirements

The successful accomplishment of any operation in which nonlethal measures are employed requires extensive preparation, of which individual, unit, medical support, and unit training are vital parts. Training should be designed to give individuals an understanding of the entire subject area and enable them to function efficiently as members of a unit. This training must be intensive and realistic. Training with the NLW capability set, and other NLW, is critical to employ these weapons with confidence. The training flow should be—train leaders, train instructors, train unit, exercise. Qualified instructors should conduct initial unit training. Unit sustainment training should be conducted per individual service and unit requirements.

2. Training of Instructors

a. Verbal communication skills, open-hand control techniques, RCA, effects of impact weapons, and civil disturbance tactics, techniques, and procedures (TTP) are areas trained to standard at the Interservice Nonlethal Individual Weapons Instructor Course (INIWIC) held at the Marine Corps Detachment, Fort Leonard Wood, Missouri. INIWIC is an Interservice Training Review Organization (ITRO) course, and quotas are available to any service through normal training reservation channels. A course outline is provided in Appendix C. This is the only formal DOD NLW instructor-training course. The training requirements outlined in the INIWIC represent the minimum training standards to include in developing NLW and tactics instructors. Additionally, the United States Army Military Police School (USAMPS) has developed a training support package that includes Army-specific TTP for small units (company level and below) in using NLW capabilities. The reference for this is the U.S. Army's website at: www.wood.army.mil.

b. The INIWIC is a "train-the-trainer" program to provide commanders NLW instructors within their command. Furthermore, the military services are not precluded from developing their own training based upon the INIWIC curriculum. The commander can then tailor the NLW instruction to fit mission needs. Establishing the service-level INIWIC serves as an assurance measure, giving the commander a quantifiable means of certifying his nonlethal instructors.

3. Nonlethal Training Topics for Deploying Units

Each service and unit will have unique training requirements based on mission type, unit size, deployment area, etc. As a minimum, the training instructor should cover the following topics in any training for units deploying with nonlethal capabilities. They are part of the INIWIC curriculum.

NOTE: U.S. Army training standards are contained in Army Regulation (AR) 350-38, Training Devices Policies and Management, 15 October 1993; USMC training standards for NLW are contained in MCO 1510.112, Individual Training Standards for Nonlethal Weapons, 27 May 1998.

---

a. Force Continuum. This is an introduction to the federal force continuum model as outlined by MCO 5500.6F, Arming of Law Enforcement and Security Personnel and the Use of Deadly Force; AR 190-14; AFI 31-207, Arming and Use of Force by Air Force Personnel (1 September 1999) (or other service orders). Identifying the proper levels of force, understanding levels of resistance, and how nonlethal technologies affect the force continuum are the minimum topics the instructor should cover.

b. Crowd Dynamics/Crowd Control. This is an introduction to the differences between crowds, mobs, and riots. This course teaches the student basic crowd control techniques that will be easily applied to various situations. The student should be familiarized with tactics and techniques found in field manual (FM) 19-15, MCWP 3-11.1, Rifle Company, TC 90-1, and Air Force manual (AFMAN) 31-201V6, Civil Disturbance (1 April 2002), but will also consider nontraditional and small unit application.

c. Communication Skills. This is an introduction to the stages of conflict management, verbal aggression, nonverbal communication, physical aggression, physiological diversions, and proper mental conditioning. The first NLW skill an individual must have is the ability to communicate. General A. M. Gray, former Commandant of the Marine Corps, once said, "A warrior's most formidable weapon is his mind." The results of an individual's mental processes are evident through actions and words, and the words and communication skills used in dealing with people are truly the first line of nonlethal options.

d. Oleoresin Capsicum (OC) Aerosol and other RCA Training. This is an introduction to the uses of OC/pepper spray and other RCA. The student should have an appreciation for decontamination requirements, storage, and shelf life of OC/RCA. The instructor will also teach legal and tactical considerations. INIWIC certified instructors should teach existing training programs that include live agent training.

e. Open-Hand Control. This is an introduction to pressure point control techniques, unarmed self-defense measures, weapon retention techniques, and other submission/restraint/search techniques. The instructor should cover handcuffing/flex-cuffing for contact team members.

NOTE: USMC plans to replace the current open-hand defensive tactics training with the USMC Martial Arts Program.

---

f. Impact Weapons. This is an introduction to using the rigid straight baton, collapsible straight batons, side handle batons, and/or riot control batons.

g. ROE/Law of Armed Conflict (LOAC). This is an introduction to the ROE/LOAC and their relationship. (Refer to DODD 5100.77, AR 27-1, and NWP 1-14M).

h. Nonlethal Munitions—Employment and Tactics. This subcourse is an introduction to the capabilities and employment of available NLW. NLW provide additional unit force protection capabilities. Introduction to tactics is outlined in Chapter 5. Appendix D is a sample unit training schedule covering the topics listed above. U.S. Army Training and Doctrine Command (TRADOC) has developed training support packages that will include company level and below TTP.

4. Leader Training

a. An understanding of the topics covered in the INIWIC curriculum is essential. However, commanders/leaders should also possess an in-depth understanding of the ROE, their implementation, as well as the politics, culture, and demographics of the area to which they are deploying.

b. Commanders should involve their medical personnel, public affairs (PA) representatives, civil affairs (CA), psychological operations (PSYOP), staff judge advocate

(SJA), engineers, explosive ordnance disposal (EOD), and military police (MP) in training exercises when possible. Involvement by these special staff sections will greatly enhance the quality and realism of the training. Medical personnel will better understand OC decontamination and the possible injuries they may encounter as a result of NLW employment. Because of the international attention that NLW receive, all personnel will benefit from media training by PA personnel. CA and PSYOP personnel can provide area-specific cultural, ethnic, and political briefs to enhance awareness of the challenges within the AO. SJA personnel can clarify complex ROE questions and assist in training on ROE-related issues. Additionally, supply and logistic staff members also require training in NLW maintenance and environmental issues unique to the equipment and munitions.

5. Unit Training

The INIWIC is primarily designed to produce instructors who will train individuals in using and employing NLW. Collective unit tactical training is the unit's responsibility; however, a NLW sample schedule based on INIWIC training is provided in Appendix D. Units should conduct collective training before deploying with NLW. Unit standard operating procedures (SOP) and tactics may require adjustment to enable the employment of NLW and new equipment for the success of specific missions. All collective training should include nonlethal options and decision making. Whenever NLW are employed during training, the instructor must also conduct lethal capability training. Troops must understand the integration of nonlethal and lethal capabilities and always be prepared to protect themselves, as required, with lethal force. It is impossible to accurately predict the response of an individual or crowd.

6. Training Disclaimer

Initial unit training should not be conducted without qualified instructors. Non-type classified and COTS nonlethal munitions cannot be fired without Army Materiel Command (AMC), Army Training Support Center (ATSC), Crane Naval Weapons Center, or Air Force Munitions Center type classifying the ammunition. The instructor should conduct unit sustainment training in accordance with (IAW) service guidelines.

7. Additional Training

a. Additional training is left to the discretion of the unit but should be encouraged in order to add credibility to the training package. Some recommended training certifications for instructors include, but are not limited to—

(1) Management of Aggressive Behavior (MOAB)
R.E.B. Training International, Inc.
PO Box 845, Stoddard, NH 03464

(2) Oleoresin Capsicum Aerosol Training (OCAT)
R.E.B. Training International, Inc.
PO Box 845, Stoddard, NH 03464

(3) Monadnock Defensive Tactics System (MDTS)
Training calendar available at the Monadnock website:
http://www.batons.com/traincal.html

b. These courses add viable credentials to the instructors as perceived by civilian and military alike. In addition, the instructor can issue certification cards for the area which an individual has been certified. Certifications in various areas offers additional protection in the event of a lawsuit or claim against the government by ensuring the individuals that have been trained within your unit are completely capable.

8. Summary

NLW and tactics training are not meant to be stand-alone training—units must anticipate the requirement for NLW in their missions and train accordingly. NLW provide commanders with a graduated response in the force continuum.

Chapter V
EMPLOYMENT

1. Background

a. A military unit deployed to an area will normally conduct operations IAW current doctrine. The environment may contain both noncombatants and combatants. Nonlethal capabilities provide the military force with tools to engage adversaries and minimize collateral damage. Fundamental to employing NLW is a thorough understanding of the force continuum. As stated before, nonlethal capabilities provide a wider range of options that augment traditional means of deadly force but do not replace them. Nonlethal capabilities allow a commander the flexibility to increase and decrease the amount of force applied to accomplish a mission. The force continuum is generally continuous and seamless, yet a careful examination reveals five broad categories:

(1) Threats. The force continuum begins with threats. These can be either implied or expressed. An implied threat may be manifested through presence by forming into riot-control formations, fixing bayonets, or donning protective masks. An expressed threat occurs when a commander makes known the consequences of defiance.

(2) Denial options. The equipment used in denial options are usually placed in highly visible areas and often marked with placards. Examples include concertina, caltrops, barbed wire, or other obstacles.

(3) Munitions that cause physical discomfort/incapacitate. Munitions that cause physical discomfort, but fall short of inflicting trauma, include flash bangs, RCA—tear gas (O-Chlorobenzyl-Malononitrile [CS]), and OC. Although the discomfort or injury may be substantially less than impact munitions, employing these munitions requires an appropriate use of force assessment. Factors such as training, discipline, prejudices, emotions, and judgment all play a part in their application.

(4) Munitions that inflict blunt trauma. Munitions that inflict blunt trauma will inflict painful or debilitating injuries. These munitions constitute the upper end of nonlethal options within the force continuum. Examples include foam batons, stingball grenades, beanbags, sponge projectiles, and rubber ball rounds.

(5) Lethal munitions. Lethal weapons are at the highest level of force continuum. Lethal options should always be regarded as part of the force continuum and not as a separate option altogether. ROE, along with judgment and experience, helps on-scene commanders decide when using lethal force is authorized and appropriate.

b. Nonlethal capabilities include using existing systems that were not designed specifically as NLW. Nonlethal capabilities can minimize fatalities, permanent injury to personnel, and undesired damage to property, facilities, and the environment. Examples include PSYOP, CA, military working dogs, and PA. However, this chapter limits discussion primarily to using weapons and components explicitly designed as part of a NLW capability set. This focus is not intended to limit the commander's option to use whatever means are available to accomplish the mission without unnecessary lethality.

2. Mission Planning Factors

Incorporating NLW into a plan does not change the fundamental principles of mission planning. Commanders should use NLW as an operational enhancement. The first concern of a commander is successfully completing the mission. ROE are a critical consideration in the commander's planning and guide him in determining the appropriate use of nonlethal and lethal force. The availability of NLW does not relieve a force from its obligation to defend itself. Therefore, it must be equipped and trained with both lethal and nonlethal means.

3. NLW Tasks

a. NLW tasks were derived from the combatant commanders and service mission needs statements and are codified in the JMAA for Nonlethal Weapons. These tasks are grouped under three core capabilities: counterpersonnel, countermateriel, and countercapability.

(1) Counterpersonnel:

(a) Control crowds.

(b) Incapacitate personnel.

(c) Deny an area to personnel.

(d) Clear facilities/structures of personnel.

(2) Countermateriel:

(a) Deny an area to vehicles, vessels, and aircraft.

(b) Disable/neutralize vehicles, vessels, aircraft, and equipment.

(3) Countercapability:

(a) Disable or neutralize facilities and systems.

(b) Deny use of WMD.

b. Although not all encompassing, the following list of employment considerations is designed to assist commanders and staffs in formulating plans for use of NLW:

(1) General considerations:

(a) Understand and define ROE.

(b) Commander develops identification of friend or foe (IFF) procedures.

(c) Distribute NLW munitions to troops in advance.

(d) NLW options require lethal overwatch fire.

(e) Ensure area is well lighted to include perimeter.

(f) Employ night vision devices outside the perimeter to identify personnel.

(g) Utilize appropriate media to inform civilians to stay away from denied areas.

(h) Enforce positive access control into perimeter (personnel/vehicles).

(i) Ensure posted signs are easily understood (symbols/pictures). If interpreter support available, post warning signs in native language.

(j) Use entanglements such as caltrops, tanglefoot, barbed wire, and concertina/razor tape as these disrupt/slow intrusion attempts.

(k) Employ sensors, if available.

(l) Employ aerial or naval pickets.

(2) Counterpersonnel considerations:

(a) Employing kinetic rounds at less than 15 feet/5 meters may result in fatal outcomes.

(b) Target area at 15 to 40 feet should be center mass for blunt munitions.

(c) Head shots may be lethal.

(d) Kinetic rounds will not be skip fired.

(e) Kinetic rounds are effective against selected individual targets and can be used to disperse individuals.

(f) Considering differing rates of fires; the M203 and shotgun are complementary in a nonlethal role, depending upon the munitions used and their effective ranges.

(g) Employ riot batons only after proper training.

(h) Use RCA to disrupt/disperse crowds.

(i) Use RCA markers (RCA with a nontoxic, water-soluble marking dye) to mark agitators for future identification and apprehension.

(j) RCA greatly reduce visibility of control force.

(k) Closely control the employment of RCA.

(l) Disperse RCA to face and eyes for full effect.

(m) Employ optics with designated marksmen to identify mob leaders.

(n) Searchlights dazzle/disorient individuals (night only).

(o) Use bullhorn for communicating with crowd and to control formations.

(p) Employ recovery/apprehension/snatch teams to apprehend agitators.

(q) Use flex cuffs for agitators.

(r) Prepare disposition plan for cleared personnel.

(s) Isolate area to prevent reinforcements.

(t) Consider future friendly use of cleared facility.

(u) Prepare security plan for cleared facility.

(v) Channel passive traffic away from formations.

(w) Maintain rapid reaction force/quick reaction force (RRF/QRF).

(3) Countermateriel/countercapability considerations:

(a) Use marking agents to mark a vessel for later interception.

(b) Consider materiel handling equipment required to evacuate vessel/airframe.

c. Selected NLW tasks are further addressed in Tables V-1 through V-7. The tables are not intended to be all-inclusive but merely a stimulus to planning. The equipment listed is taken from the currently available systems designed for nonlethal purposes. The user should not limit his nonlethal options to this specific equipment and should be aware that there are many existing and emerging systems and potential uses of current technology and common objects to achieve the desired mission end state.

Table V-1. Crowd Control

Tasks	*NLW Planning Factors*	*Equipment*	*Employment Considerations*	*Notes*
Crowd control	Administration/ Personnel • SJA / claims planning • Chaplain affairs • Casualty notification • Detainee disposition plan • PAO/media planning • Information dissemination	Support Equipment • Riot face shield • Full length riot shield • Expandable riot baton • Optics • Portable bullhorn • Vest	• Distribute to troops in advance • Employ riot baton only after proper training • Employ optics with designated marksman (DM) to identify (ID) mob leaders • Use bullhorn for communicating with crowd and to control formations	• Strictly define ROE • NLW options require lethal cover fire
Crowd control	Intelligence • Intelligence preparation of the battlespace (IPB) • Background on country and culture • Route/map recon • Sensor emplacement for advance notification • Information on Mob leaders • Patrols and listening post (LP)/observation post (OP) • Determine reason for the crowd	Kinetic Rounds • 12-gauge (ga) bean bag round • 12-gauge rubber bullet • 12-gauge launching cartridge • 40-mm rubber baton • 40-mm wooden baton **NOTE:** USMC no longer uses wooden baton. • 40-mm stinger grenade • Stun grenade • Flash bang	• Do not employ kinetic rounds at less than 15 feet due to possible fatal outcome • Target area at 15 to 40 feet should be center mass • Head shots are not acceptable • Kinetic rounds will not be skip fired • Kinetic rounds effective against selected targets and to disperse individuals • M203 and shotgun are nonlethal complementary; M203 has low rate of fire; area target. Shotgun has high rate of fire; point target	

Table V-1. Crowd Control

Tasks	***NLW Planning Factors***	***Equipment***	***Employment Considerations***	***Notes***
Crowd control	<u>Operations</u> • Tactical Integration of NLW and lethal fires • Fire Support Element (FSE) plans to include RCA • Allocation of NLW munitions • Integration of NLW fires and maneuver • Engineer barrier planning • Understand mission (deny access/disperse/ monitor)	<u>RCA</u> • OC Dispenser • Team OC dispenser • High volume dispenser	• Disrupt/disperse crowds • Disperse RCA to face and eyes for full effect • Employ snatch teams to apprehend agitators	
Crowd control	<u>Logistics</u> • Distribution of NLW supplies • Security of rear area facilities • Resupply of NLW munitions • Procurement of medical supplies • Hospital planning • Medic planning for tactical units	<u>Riot Control</u> • Xenon Searchlight • Flex cuffs • RCA markers	• Searchlights dazzle/disorient individuals (night only) • Flex cuffs used for agitators • RCA markers are used to mark agitators for future ID	

Table V-2. Incapacitate Personnel

Task	*NLW Planning Factors*	*Equipment*	*Employment*	*Notes*
Incapacitate personnel	Administration/Personnel • SJA/claims planning • Media plan	Support Equipment • Riot face shield • Full length riot shield • Expandable riot baton • Rifleman's combat optic • Vest	• Distribute to troops in advance • Do not employ riot baton to the head • Employ optic with DM to ID mob leaders and cover with lethal fire	• Commander needs to define incapacitation mission • Take a balanced approach to public dissemination of NLW so counter measures cannot be rapidly developed • Cover NLW options by lethal force
Incapacitate personnel	Intelligence • ID types of insurgents and demonstrators • ID causes and factions • ID mob ringleaders	Kinetic Rounds • 12-gauge bean bag round • 12-gauge rubber bullet • 12-gauge launching cartridge • 40-mm rubber baton • 40-mm wooden baton **NOTE:** USMC no longer uses wooden baton. • 40-mm stinger grenade • Stun grenade • Flash bang • M203 and shotguns are nonlethal complementary. M203 has a low rate of fire, area target. Shotgun has a high rate of fire, point target	• Do not employ kinetic rounds at less than 15 feet due to possible fatal outcome • Target area at 15-40 feet should be center mass • Head shots are not acceptable • Kinetic rounds will not be skip fired • Kinetic rounds effective against selected targets and to disperse individuals • M203 and shotguns are nonlethal complementary. M203 has a low rate of fire, area target. Shotgun has a high rate of fire, point target	

Table V-2. Incapacitate Personnel

Task	*NLW Planning Factors*	*Equipment*	*Employment*	*Notes*
Incapacitate personnel	Operations • Prepare and confirm ROE for NLW • Integration of NLW and deadly force • Allocate NLW and designate unit use • Integrate NLW fires and maneuver	RCA • OC dispenser • Team OC dispenser • High-volume • OC dispenser	• Disrupt/disperse crowds • Disperse RCA to face and eyes for full effect	
Incapacitate personnel	Logistics • Distribute NLW munitions • ID special storage requirements • Resupply NLW munitions	Riot Control • Xenon • searchlight • Flex cuffs • Chemical markers	• Searchlights dazzle/disorient individuals (night only) • Use flex cuffs for agitators • Use RCA markers to mark agitators for incapacitation	
Incapacitate personnel	Civil Affairs • Collect civilian intelligence • Care for injured personnel • Detain civilian personnel			

Table V-3. Deny Area to Personnel

Task	*NLW Planning Factors*	*Equipment*	*Employment*	*Notes*
Deny area to personnel	Administration/ Personnel • SJA/claims planning • Media plan	Support Equipment • Area lights • Area sensors • Night vision • Access control • Warning signs written in host nation language	• Ensure area is well lighted to include perimeter • Employ night vision devices outside the perimeter • Positively ID accessed personnel • Employ warning signs in native language	• Utilize appropriate media to warn civilians away from denied area

Table V-3. Deny Area to Personnel

Task	*NLW Planning Factors*	*Equipment*	*Employment*	*Notes*
Deny area to personnel	Intelligence • Terrain and map study • Threat ID • Sensor emplacement around area • IPB			
Deny area to personnel	Operations • Prepare and confirm ROE • Integrate NLW and deadly force • Allocate NLW and designate unit use • Determine scope of mission • Determine access procedures • Use of area (friendly or total denial)			
Deny area to personnel	Logistics • Distribute NLW munitions • ID special storage requirements • Resupply NLW munitions • Security of rear area facilities.			
Deny area to personnel	Civil Affairs • Collect civilian intelligence	Entanglements • Caltrops • Tanglefoot • Barbed wire • Foams • Concertina/ razor tape • Stinger spike system	• Disrupt/slow intrusion attempts • Maintain suitable reaction force • Channel passive traffic away • Use foam to cover caltrops	

Table V-4. Deny Area to Vehicles

Task	*NLW Planning Factors*	*Equipment*	*Employment*	*Notes*
Deny area to vehicles	Administration/ Personnel • SJA/claims planning • Media plan	Support Equipment • Area lights • Area sensors • Night vision • Access control • Warning signs written in host nation language	• Ensure area is well lighted to include perimeter • Employ night vision devices outside the perimeter • Positively ID accessed vehicles	• Commander needs to define denial mission • Utilize appropriate media to warn civilians away from denied area
Deny area to vehicles	Intelligence • Terrain and map study • Threat ID • Sensor emplacement around area • IPB • Traffic analysis			
Deny area to vehicles	Operations • Prepare and confirm ROE • Allocate NLW and designate unit use • Determine scope of mission • Determine access procedures • Determine MP/ security forces requirements • Use of area (friendly or total denial)			• Commanders must consider flying operations when employing RCA. Area(s) should be reusable in short order.
Deny area to vehicles	Logistics • Distribute NLW supplies • Be prepared to evacuate stalled vehicles			
Deny area to vehicles	Civil Affairs • Collection of civilian intelligence	Entanglements • Caltrops • Tanglefoot • Barriers • Organic equipment • Barbed wire • Foams	• Disrupt/slow intrusion attempts • Maintain suitable reaction force • Channel passive traffic away • Use foam to cover caltrops	• Commanders must consider flying operations when employing NLW. NLW should not hinder flying operations.

Table V-5. Deny/Neutralize Vehicles, Aircraft, Vessels, and Facilities

Task	*NLW Planning Factors*	*Equipment*	*Employment*	*Notes*
Deny/ neutralize vehicles, aircraft, vessels, facilities	Administration/Personnel • SJA/claims planning • Media plan	Support Equipment • Sensors to track approaching vehicles, aircraft, and vessels • Night vision equipment • Materiel handling equipment • Portable bullhorn	• Employ ground and water sensors, if required • Employ aerial or naval pickets • Use materiel handling equipment to evacuate vessel	• Commander needs to define neutralization mission • Element of surprise remains a critical factor in mission accomplishment • Commander needs to develop IFF procedures
Deny/ neutralize vehicles, aircraft, vessels, facilities	Intelligence • Sensor emplacement if required • Threat ID • IPB • Analyze target area			
Deny/neutralize vehicles, aircraft, vessels, facilities	Operations • Prepare and confirm ROE • Integrate NLW and deadly force • Allocate NLW and designate unit use • Determine scope of mission • Determine access procedures • Determine boarding procedures (vessels and aircraft) • Determine apprehension procedures • Determine decontamination (DECON) procedures	RCA • Marking agent	• Use marking agents to mark a vessel for later interception	

Table V-5. Deny/Neutralize Vehicles, Aircraft, Vessels, and Facilities

Task	*NLW Planning Factors*	*Equipment*	*Employment*	*Notes*
Deny / neutralize vehicles, aircraft, vessels, facilities	Logistics • Distribute NLW munitions • ID special storage requirements • Resupply NLW munitions • Determine vessel/aircraft disposition	Riot Control • Xenon searchlight	• Use RCA to incapacitate crew members • Searchlights can disorient personnel or illuminate target	• Commanders must consider flying operations when employing RCA • Area(s) should be reusable in short order.
Deny / neutralize vehicles, aircraft, vessels, facilities	Civil Affairs • Collect civilian intelligence • Handle displaced civilian personnel	Entanglements • Caltrops • Foams • Speed bumps • Nets • PVAB	• Use caltrops to disable vehicle • Use combination of foam and caltrops to preserve the element of surprise • Channel passive traffic away • Use nets to foul props	
Deny/neutralize vehicles, aircraft, vessels, facilities	Medical • Treating injured civilian personnel			

Table V-6. Seize Personnel

Task	*NLW Planning Factors*	*Equipment*	*Employment*	*Notes*
Seize personnel	Administration/ Personnel • SJA/claims planning • PAO media plan	Support Equipment • Optics	• Employ optics with DM to ID personnel to be seized • Cover personnel employing nonlethal force	• Commander needs to define seizure mission and strictly enforce ROE • Element of surprise is critical for mission accomplishment • Speed is critical • Seizing personnel can be surgical with special operations personnel or conducted within a crowd to seize an agitator
Seize personnel	Operations • Prepare and confirm ROE • Integrate NLW and deadly force • Allocate NLW and designate unit use • Integrate NLW fires and maneuver • Breaching and assault planning • Determine holding area/processing center	RCA • OC dispenser • CS	• Use RCA to incapacitate personnel • Spray into eyes and face for full effect	
Seize personnel	Logistics • Distribute NLW munitions • ID special storage requirements • Resupply NLW munitions • Plan transportation	Riot Control • Xenon searchlight • Flex cuffs • RCA markers	• Searchlights dazzle/disorient individuals • Use flex cuffs for apprehension • Use RCA markers to mark agitators for seizure	

Table V-6. Seize Personnel

Task	*NLW Planning Factors*	*Equipment*	*Employment*	*Notes*
Seize personnel	Civil Affairs • Collect civilian intelligence • Care for injured personnel • Detain civilian personnel			
Seize personnel	Medical • Medical evacuation (MEDEVAC) planning			

Table V-7. Clear Facilities of Personnel

Tasks	*NLW Planning Factors*	*Equipment*	*Employment Considerations*	*Notes*
Clear facilities of personnel	Administration/ Personnel • SJA/claims planning • Chaplain affairs • Casualty notification • PAO media plan	Support Equipment • Riot face shield • Full-length riot shield • Expandable riot baton • Optics • Portable bullhorn	• Distribute to troops in advance • Do not employ riot baton to the head • Employ optic with DM to ID occupant leaders and provide lethal cover fire • Use bullhorn used for communicating with occupants	• Surprise is a critical element for mission accomplishment • Commander must strictly define ROE • Due to close quarter tactics, the force must be instantly ready to utilize deadly force if required

Table V-7. Clear Facilities of Personnel

Tasks	*NLW Planning Factors*	*Equipment*	*Employment Considerations*	*Notes*
Clear facilities of personnel	Intelligence • IPB • Background on facility, access, power plant • Route/map recon • Background info on building occupants • Determine any outside support for building occupants	Kinetic Rounds • 12-gauge bean bag round • 12-gauge rubber bullet • 12-gauge launching cartridge • 40-mm rubber baton • 40-mm wooden baton **NOTE:** USMC no longer uses wooden baton. • 40-mm stinger Stun grenade • Flash bang	• Do not employ kinetic rounds at less than 15 feet due to possible fatal outcome • Targeted personnel at 15 to 40 feet should be center mass • Head shots are not acceptable • Kinetic rounds will not be skip fired • Kinetic rounds effective against selected targets and to disperse individuals • Stun grenade effects can be amplified in buildings • Visibility will be reduced once kinetic rounds are employed • M203 and shotguns are nonlethal complementary. M203 has a low rate of fire, area target. Shotgun has a high rate of fire, point target	• Using RCA greatly reduces visibility • Use of RCA is restricted to certain circumstances; verify that proper authority has been given before use

Table V-7. Clear Facilities of Personnel

Tasks	*NLW Planning Factors*	*Equipment*	*Employment Considerations*	*Notes*
Clear facilities of personnel	Operations • Tactically integrate NLW and lethal fires • FSE plans to include RCA • Allocate NLW munitions • Integrate NLW fires and maneuver • Determine assault and breaching plan • Plan for isolation of facility	RCA • OC dispenser • Team OC dispenser • High-volume dispenser	• Disrupt/disperse crowds • Apply RCA to face and eyes for full effect	• Isolate area to prevent reinforcements • Prepare disposition plan for cleared personnel • Consider use of facility • Prepare security plan for cleared facility
Clear facilities of personnel	Logistics • Distribute NLW supplies • Resupply NLW munitions • Plan for transportation of displaced civilians • Plan facility maintenance or repair	Riot Control • Xenon searchlight • Flex cuffs • RCA markers • Flash bangs	• Searchlights dazzle/disorient individuals • Use flex cuffs for agitators • Use RCA markers to mark agitators for future ID • Flash bangs to dazzle/disorient individuals	
Clear facilities of personnel	Civil Affairs • Collect civilian intelligence • Care for injured civilians • Coordinate plans for controlling civilians			
Clear facilities of personnel	Medical • Hospital planning • Medic planning for tactical units • Procurement of medical supplies			

4. Staff Planning/Responsibilities

a. Administration. A staff composed of the G-1/S-1, PAO, SJA, and others as needed, is responsible for advising the commander on NLW employment and ROE. Additionally, administrative personnel are responsible for processing all civilian detainees and claims that arise from NLW employment. As a minimum, the staff must consider the following:

(1) SJA/claims planning.

(2) Chaplain.

(3) Casualty notification.

(4) Detainee disposition plan.

(5) PAO/media planning.

(6) Information dissemination.

b. Intelligence. The G-2/S-2 office is responsible for conducting the IPB process and providing information on the adversary's culture, political stability, and level of dedication to the cause, motivation, leadership, capabilities, equipment, and tactics. The G-2/S-2 should locate native language speakers to assist the interrogator-translator team.

(1) Background on country and culture.

(2) Route/map reconnaissance.

(3) Place sensors for advance notification.

(4) Information on leaders.

(5) LP/OP.

(6) Determine reason for the crowd.

(7) Identify threat.

(8) Analyze traffic.

c. Operations. The G-3/S-3 office is responsible for unit training, nonlethal certification of units, mission analysis, maneuver plans, and CA planning when required.

(1) Tactically integrate NLW and lethal fires.

(2) Fire support plans to include using RCA.

(3) Allocate NLW munitions.

(4) Plan engineer barrier.

d. Logistics. The G-4/S-4 office is responsible for logistics considerations, ship-to-shore movement of weapons and ammunition, storage and maintenance of equipment, resupply, transportation ashore, and other support requirements inherent in using NLW.

(1) Distribute NLW supplies.

(2) Security of rear area facilities.

(3) Resupply NLW munitions.

(4) Special maintenance requirements for NLW equipment and munitions.

e. Communications Officer. The G-6/S-6 staff is responsible for installing, operating, and maintaining communications from units confronting belligerents to the tactical operation center (TOC) or combat operations center (COC). The staff must plan for the effects NLW will have on communications to include proper frequencies.

f. PAO. The PAO provides liaison with outside media and participates in the staff planning process. PAO is also responsible for ensuring the NLW media plan and news releases consider the perceptions of the native media and belligerents or adversaries. PA personnel must receive training on NLW technologies, employment and capabilities, acute and long-term effects, impact on the environment, and legal approval procedures (at a minimum) so they can respond to media and public inquiries.

g. Medical. Medical personnel must be aware of the effects of NLW. Medical personnel should also be familiar with and able to provide decontamination of RCA as well as administer antidotes in cases of adverse reactions to riot control and other agents.

(1) MEDEVAC planning.

(2) Procuring medical supplies.

(3) Hospital planning.

(4) Medical planning for tactical units.

h. Chaplain. The commander should consider chaplain support for counseling and/or religious services for military and civilian personnel. Additionally, chaplains should provide information regarding the religious and cultural background of the local populace.

i. CA. The CA office is responsible for planning and coordinating all aspects of CA activities and providing assistance in gathering intelligence from civilian sources.

(1) Handle displaced civilian personnel.

(2) Coordinate plans for control of civilians.

j. PSYOP. Tactical PSYOP forces serve as an ideal tool for the commander to resolve issues of armed resistance and local interference with military operations while encouraging dialogue and cooperation with noncombatants. This force provides dissemination of PSYOP products (loudspeaker messages, leaflets, posters, etc.) and conducts face-to-face communications with target audience. PSYOP forces provide the supported commanders with nonlethal leverage by influencing foreign populations and expressing information subjectively to influence attitudes and behavior, thereby obtaining compliance or noninterference. They provide public information to foreign populations to support humanitarian activities, ease suffering, and restore or maintain civil order. They serve as the supported commander's voice to foreign populations to convey intent, countering enemy propaganda, misinformation, and opposing information to correctly portray friendly intent and actions. Finally, PSYOP forces deny others the ability to polarize public opinion and political will against the United States and its allies within an AO.

5. Tactics

The commander should include the following considerations in preparing for unit tactical operations involving NLW:

a. Avoid reconfiguring small-unit organization, except for special configurations such as—

(1) Special teams.

(2) Attachments.

(3) Crowd control formations.

(4) Security organizations.

b. Standardize postures in unit tactical SOP. Take the following steps to add a higher volume of specialty impact munitions and equipment within squads, platoons, or units:

(1) Avoid arming troops involved in apprehending designated personnel with long weapons (i.e., shotguns, rifles) because they can be easily grabbed/pulled away by individuals.

(2) Add weapons systems (shotguns) that provide a greater nonlethal capability. When loaded with nonlethal munitions, the longer reloading time of the M203 grenade launcher is protected by coverage from the high rate of fire from shotgun.

(3) Add nonlethal munitions to existing organic weapons systems (M203 grenade launcher).

(4) Avoid arming troops in the front lines of crowd control formations with long weapons.

(5) Focus riot-control capabilities in specialized reaction forces.

(6) Establish base line NLW postures.

(7) Establish command relations for special units.

(8) Maximize distance and barriers.

(9) Create NLW range cards for static positions.

(10) Provide the force with available OC/RCA consistent with ROE.

(11) Always mutually support forces utilizing NLW with lethal means.

(12) Consider environmental conditions (heat/cold/wind) and their effect on the performance of nonlethal munitions. Proper storage, periodic inspections, and rotation of NLW ammunition are critical to maintaining the effectiveness and viability of nonlethal munitions.

c. Unit Weapons, Unit Configurations, and Ammunition Considerations.

(1) Individual. Designate individuals as nonlethal shooters. Nonlethal shooters should also carry lethal munitions for use as required. Lethal rounds must be marked and carried in a manner to prevent confusing nonlethal rounds with lethal rounds.

(2) Squad. Squad leaders should dictate who carries nonlethal grenades and maintain their weapons loaded with lethal ammunition. Ideally, the squad should not change its task organization to accommodate nonlethal equipment, but should designate nonlethal shooters.

(3) Patrols. Commanders should not plan a nonlethal patrol, but should plan a combat/security patrol with a nonlethal capability when mission dictates.

(4) Static Positions. Individuals in static positions should have their weapons loaded with lethal munitions (mission, enemy, terrain and weather, troops and support available–time available and civil considerations dependent) (METT-T/METT-TC). Additionally, they should carry nonlethal munitions. Both shotgun and M203 work well from static positions and should complement each other in a nonlethal mode.

(5) Recovery Teams. Recovery teams have the primary responsibility of retrieving and controlling designated personnel from the crowd; the commander should establish a recovery team before deploying. The force should train team members in unarmed self-defense, open-hand control, and hand cuffing/flex cuffing. It should equip recovery teams with personal protection gear, 9-mm weapon, and flex cuffs. Due to the physical nature of recovery teams, it is important not to arm team members with long rifles. If security for the recovery team is an issue, the commander should attach a security element.

(6) Crowd Control Formations. The unit should establish crowd control formations within a minimum response time. Because of the physical nature of crowd control, individuals in these formations, if possible, should not carry long rifles. These formations should have inherent nonlethal capability. The unit must provide lethal coverage for the entire formation.

d. Individual Elements of a Formation. These elements are: base element, support element, and command element. The formation needs to provide versatility. Restrictions must never be set on numbers, equipment, or changes to the standard formations.

(1) Base Element. Role of the base element (front line):

(a) Base element serves as the front line or the base/foundation of the formation (can be divided into two or three sections or ranks).

(b) Front line consists of the shield holders.

(c) Angle the bottom of the shield to channel flammable liquids away from the body.

(d) Distance or interval between the shields differs based upon the range to the crowd. The closer the crowd, the tighter the interval between the shields.

(e) Provide frontal security.

(f) Hold a solid line.

(g) Provide intimidation factor.

(h) Fire team leaders are in the line holding shields.

(i) Secondary (tertiary) line(s) can consist of the baton holders and/or nonlethal gunners.

(j) Baton holders will—

- Strike with batons.
- Spray OC or disperse smoke.
- Serve as spotters for the NLW gunners.
- Control and detain subjects pulled from the crowd.

(k) Nonlethal gunners will—

- Consist of team leader with M203/M16 (provides nonlethal cover with M203 40-mm munitions and lethal cover with M16).
- Position themselves between shields working with baton holders.
- Serve as immediate control for the team.
- Maintain continuous muzzle awareness of the weapon.
- Provide immediate support to the shields.

(2) Support Element. Role and employment of the support element:

(a) Provide the means to employ lethal and nonlethal capabilities.

(b) Can consist of two ranks: M16/M203 (2nd rank) Mossberg 12-gauge shotguns (3rd rank).

(c) Provide security for search and recovery team.

(d) Consists of munitions teams and/or recovery teams.

(e) Munitions teams provide the means to employ lethal and nonlethal capabilities consisting of M16/M203 and Mossberg 12-gauge shotguns.

(f) Recovery teams provide ability to detain and search rioters downed by nonlethal munitions. Recovery teams will—

- Extract vehicles or personnel from the crowd who are in immediate danger.
- Detain and escort downed rioters to rear of the formation.
- Restrain and search detained individuals.
- Communicate with the base element.
- Spray or employ RCA as needed.
- Fill in the gaps in the front line as needed.
- Insert into a confined or congested area too large for the formation.
- Use a soft base line that capitalizes on the munitions fired to deploy forward of the shields in order to drive the crowd back, retrieve personnel/gear, and apprehend key riot personnel. Employed in pairs (1 x shield man/1 x shotgun).

NOTE: Refer to Marine Corps Security Force (MCSF) formation examples.

---

(3) Command Element. Role of the command element is to provide command and signal to the formation and consists of the platoon leader/flight commander/ platoon sergeant/flight sergeant, and squad leaders.

(a) Platoon Leader/Flight Commander, Platoon Sergeant/Flight Sergeant—

- Pass verbal commands.
- Maintain order within formations.
- Gauge and evaluate use of munitions and resupply considerations.

(b) Squad Leaders—

- Effectively control the squad.
- Pass information up the chain.
- Employ OC as directed.

(c) Designated Marksman (DM). During a nonlethal engagement, using DM provides confidence and safety to those facing a riot. The DM in an overwatch position scans the crowd, identifies threats and designated personnel for recovery, and fires lethal rounds if a lethal threat is presented. Additionally, DM are ideally suited for flank security and counter sniper operations.

6. Employment Objectives

a. Innovative employment methods of NLW and ingenuity take on critical significance. NLW are employed in concert with other nonlethal and lethal systems to create a synergistic effect.

b. The measures for countering many nonlethal options are usually apparent, quickly learned, and readily available. Because they are not intended to kill, nonlethal options "teach" an adversary what to avoid in the future. Small unit commanders must prepare to keep one step ahead of motivated belligerents.

c. Many nonlethal munitions have both maximum effective and minimum safe ranges. Individuals short of the minimum safe range may suffer severe injuries or death while the effects of most nonlethal munitions are greatly decreased at longer ranges. To be effective, an adversary must be engaged within the optimum range of the munition. Figure V-1 illustrates the ranges for a variety of nonlethal munitions and systems.

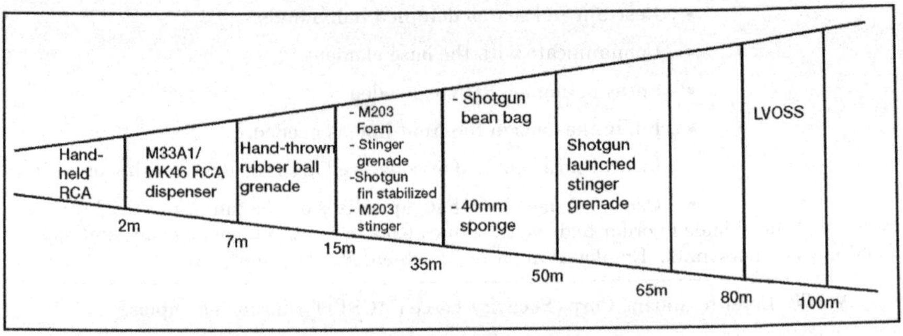

Figure V-1. Nonlethal Weapons Ranges

7. Rules of Engagement

a. U.S. forces follow directives issued by competent military authority that delineate the circumstances and limitations under which deadly force will be initiated against other forces. Intrinsic to the effective use of nonlethal options is the authorization to employ them. ROE are the primary means by which the chain of command can provide guidance to deployed forces in peacetime for handling crises and, in wartime, to help control the level of hostilities.

b. The commander should distinguish ROE from tactical control measures, threat conditions, and arming orders/weapons conditions. Although arming orders are sometimes listed on the same quick reference card as ROE for easy reference, the commander should distinguish ROE from other policies and directives developed, such as weapons confiscation rules.

c. The commander should delegate the decision to use NLW against an adversary to the lowest possible level, preferably to the platoon, flight, or squad level. However, this requires that all personnel, not just leaders, have a clear understanding of ROE and his intent.

d. Using lethal force, employed under the standing ROE, will never be denied. At no time will forces be deployed without the ability to defend themselves against a lethal threat nor will they forego normal training, arming, and equipping for combat. Nonlethal options are a complement to, not a replacement for, lethal force and seek to expand a proactive response across the range of military operations.

e. The following figures are examples of service-specific formations. These examples are not all-inclusive, and the tactical situation will always dictate actual unit employment.

FM 19-15, *Civil Disturbance*; MCWP 3-11.1, *Rifle Company;* and MCWP 3-11.2, *Marine Rifle Squad* provide extensive detail about small-unit formations, their use, and other considerations. Figures V-2 to V-7 are from I and II Marine Expeditionary Force (MEF) Special Operations Training Groups (SOTG) and represent a notional infantry platoon. Figures V-8 to V-13 represent a notional U.S. Army infantry platoon. Figures V-14 to V-21 are additional platoon formations submitted by Marine Corps Security Force Training Company.

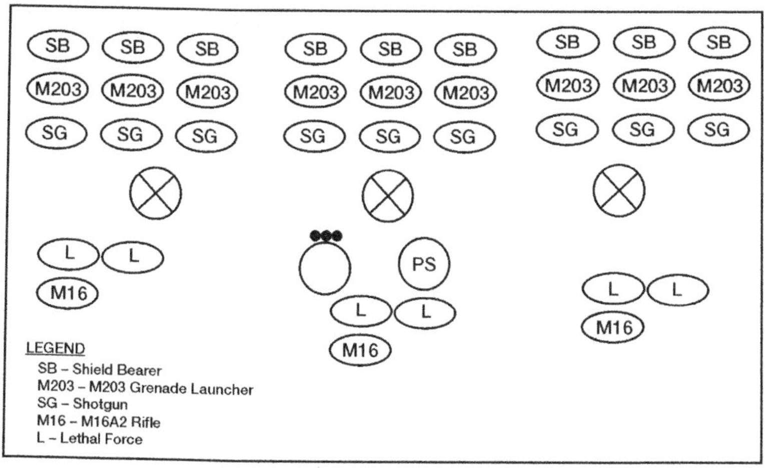

Figure V-2. Platoon in Depth

NOTE: The platoon is the smallest unit capable of independent maneuver. The platoon in depth formation shown above provides the following benefits:

- In-depth support with three ranks.
- Complementary support with the shotguns covering the slower reload time for the M203.
- Additional personnel from each squad to form the recovery teams.
- Each squad is capable of moving as an independent formation forward of the platoon line, if situation dictates.
- Other tactical formations are applicable—echelon, wedge, column.

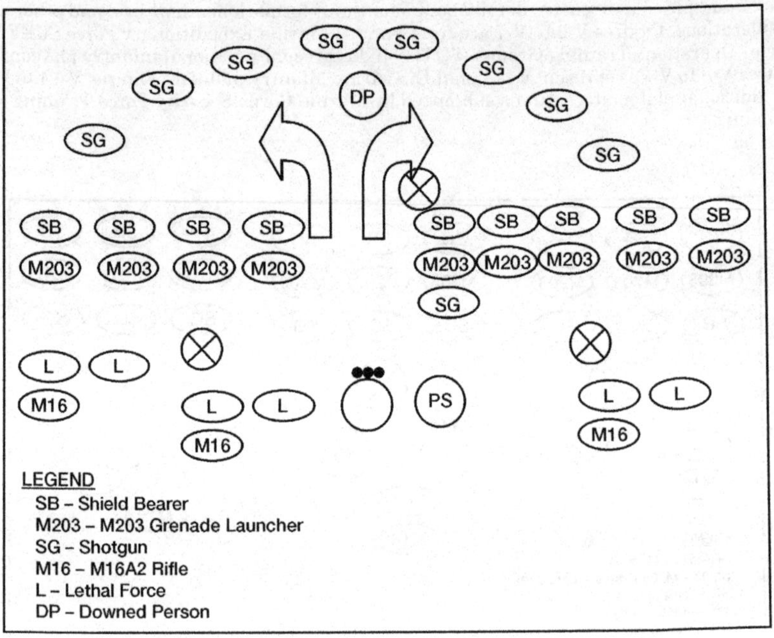

Figure V-3. Platoon Recovery Team Deployment—Security Element

---

NOTE: Security element deploys around downed person.

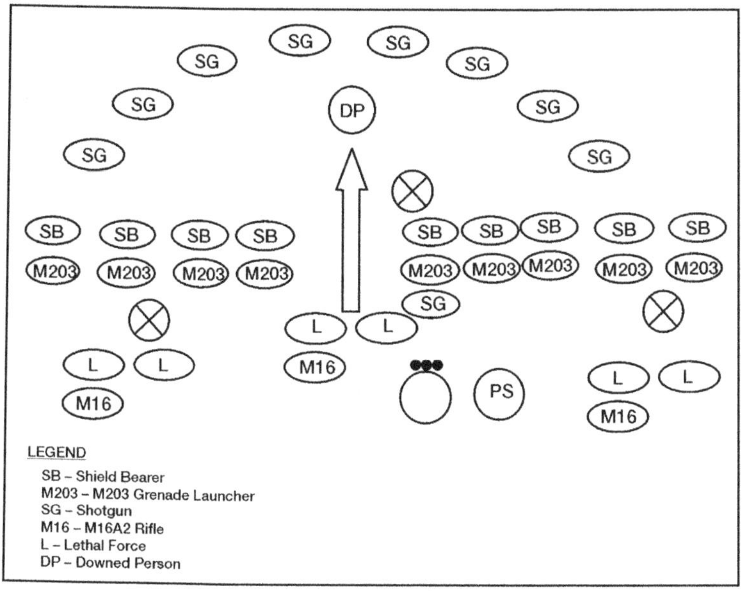

Figure V-4. Platoon Recovery Team Deployment—Recovery Team

NOTE: Recovery team deploys to detain, search, and recover downed person.

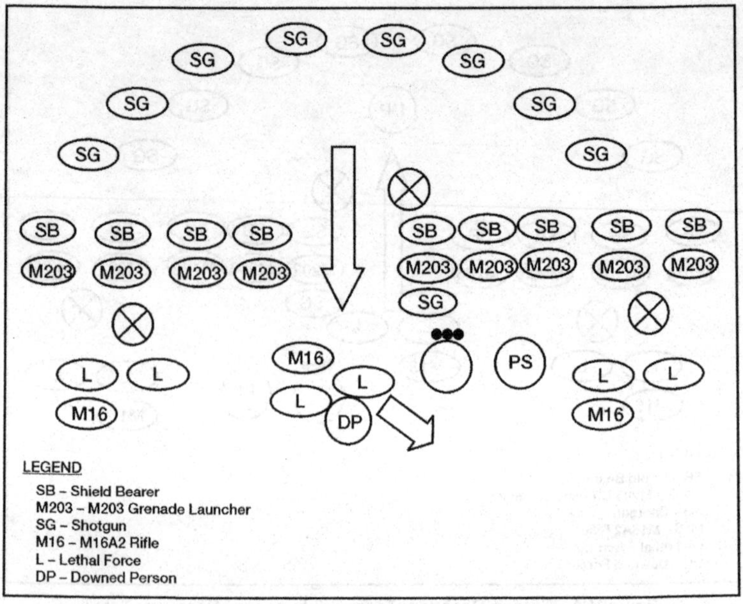

Figure V-5. Platoon Recovery Team Deployment—Recovery Team Redeploy

---

NOTE: Recovery team redeploys with detained person.

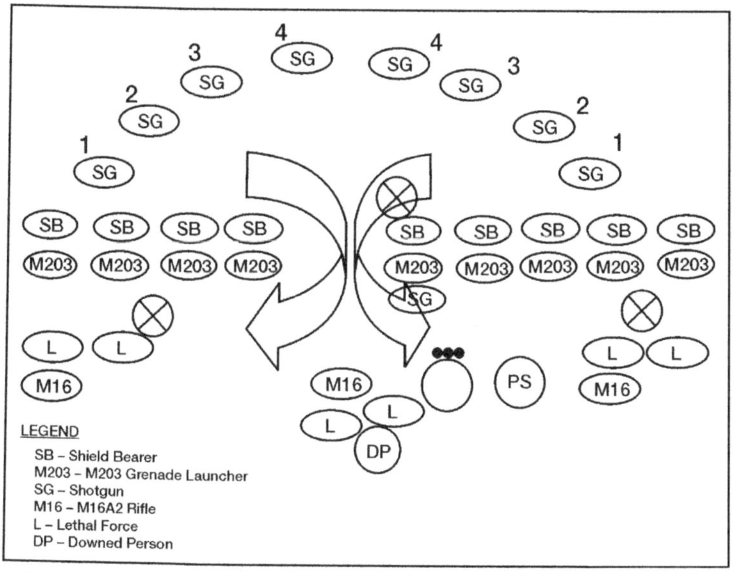

Figure V-6. Platoon Recovery Team Deployment—Security Element Redeploy

---

NOTE: Security element redeploys back to 3rd rank "by the numbers." Team leader calls "4s in," then "3s in," etc. The last person in calls "last man in" and formation closes the gap.

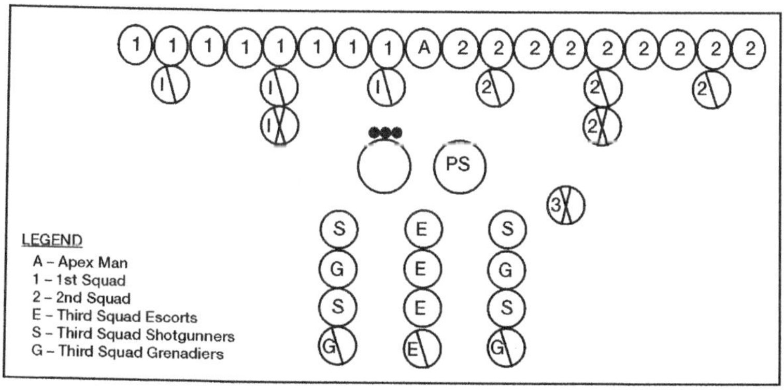

Figure V-7. Platoon On Line—Munitions Squad in Close Support to Cover Flanks

---

NOTE: Platoon in close support. NLW and lethal weapons are sited to cover the flanks of the formation, as well as support the front line.

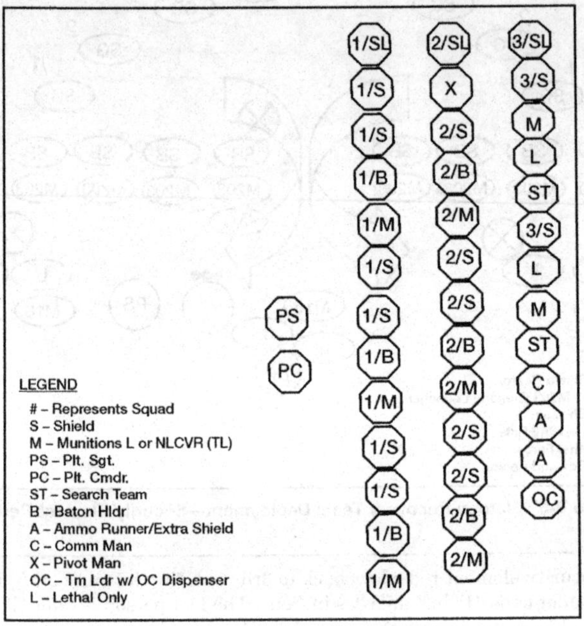

Figure V-8. Platoon Column

NOTE: Platoon in column formation. For administrative movement only.

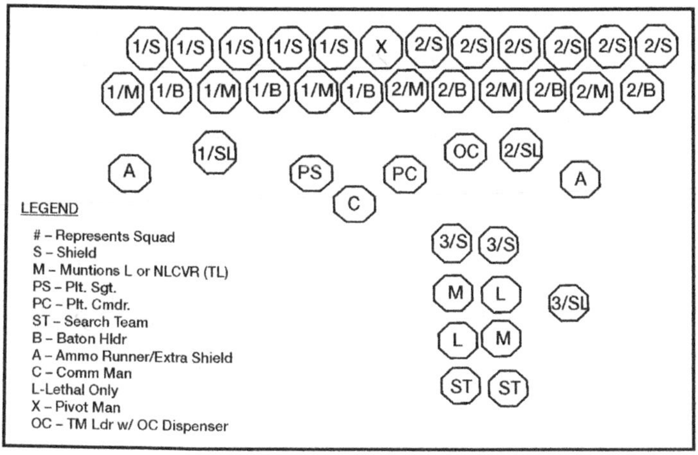

Figure V-9. Platoon On Line

NOTE: Platoon on line—basic formation for the employment of NLW.

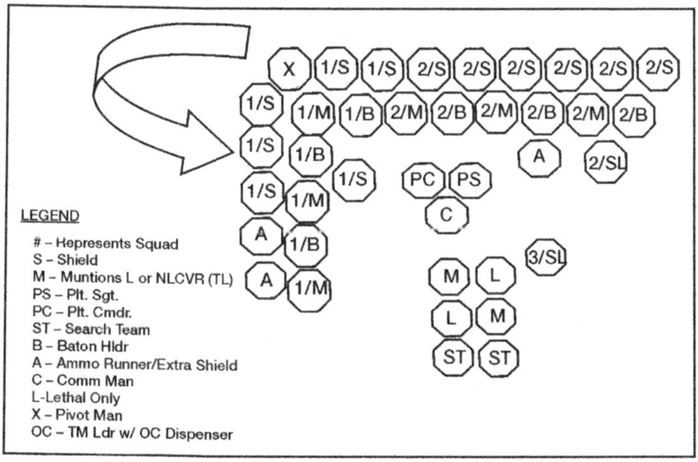

Figure V-10. Platoon Lateral Support, Left (Right)

NOTE: Platoon in lateral support to protect flanks.

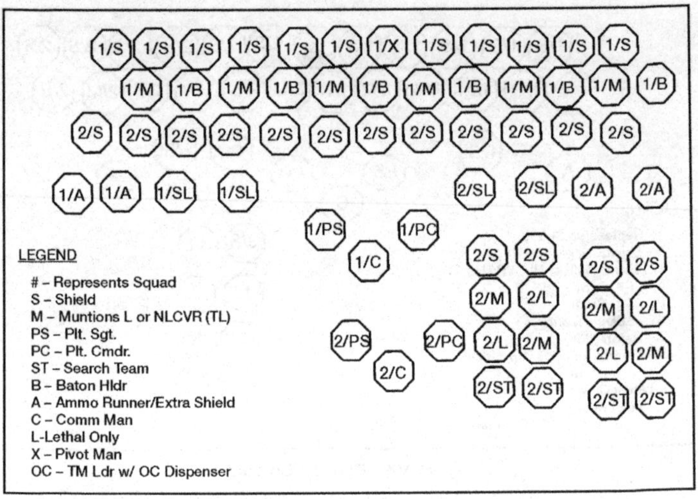

Figure V-11. Two Platoons in Close Support

NOTE: Two platoons in close support—tied in for mutual advantage.

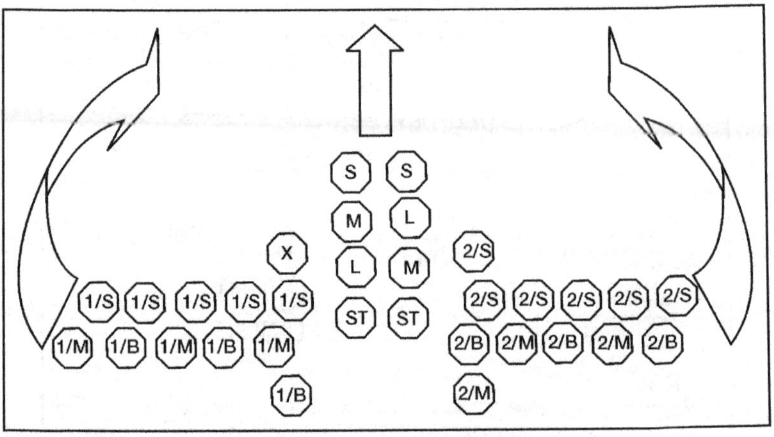

Figure V-12. Recovery Team Deployment

NOTE: Recovery team deploys from formation—remainder of platoon can deploy around the recovery team from either flank.

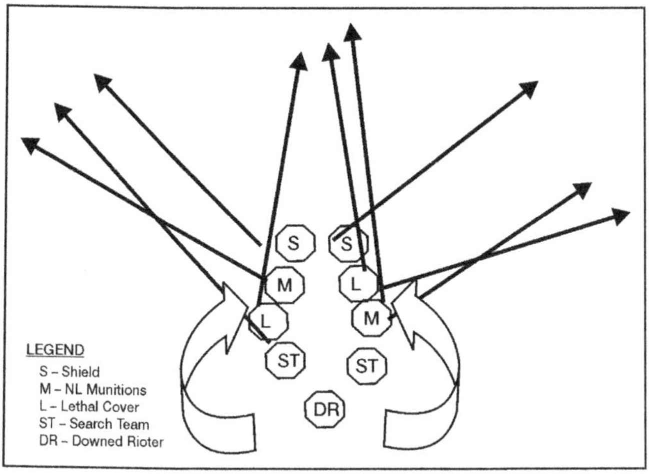

Figure V-13. Recovery Team Actions and Security Posture

NOTE: Recovery team forms security perimeter around downed rioter, conducts search and detain actions. Once actions are complete, recovery team redeploys.

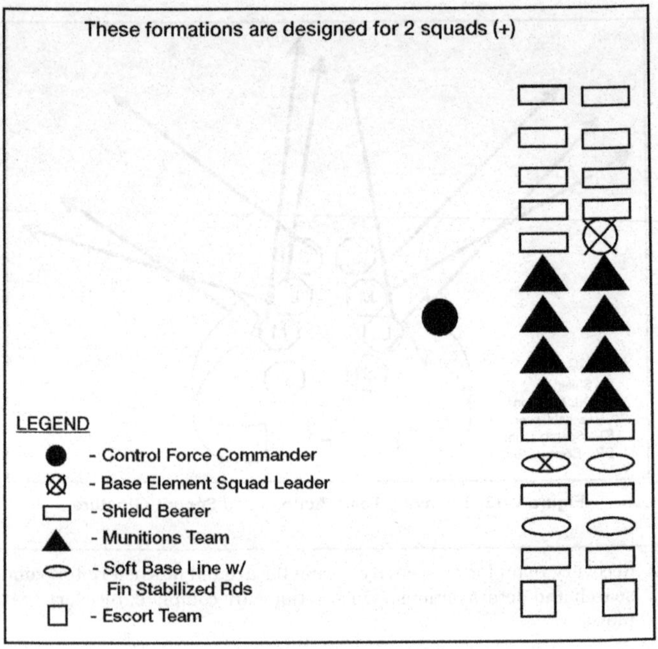

Figure V-14. Platoon in Column (USMC)

NOTE: These formations are designed for 2 squads (+). This allows the platoon to move rapidly when there is no chance of being engaged with the crowd.

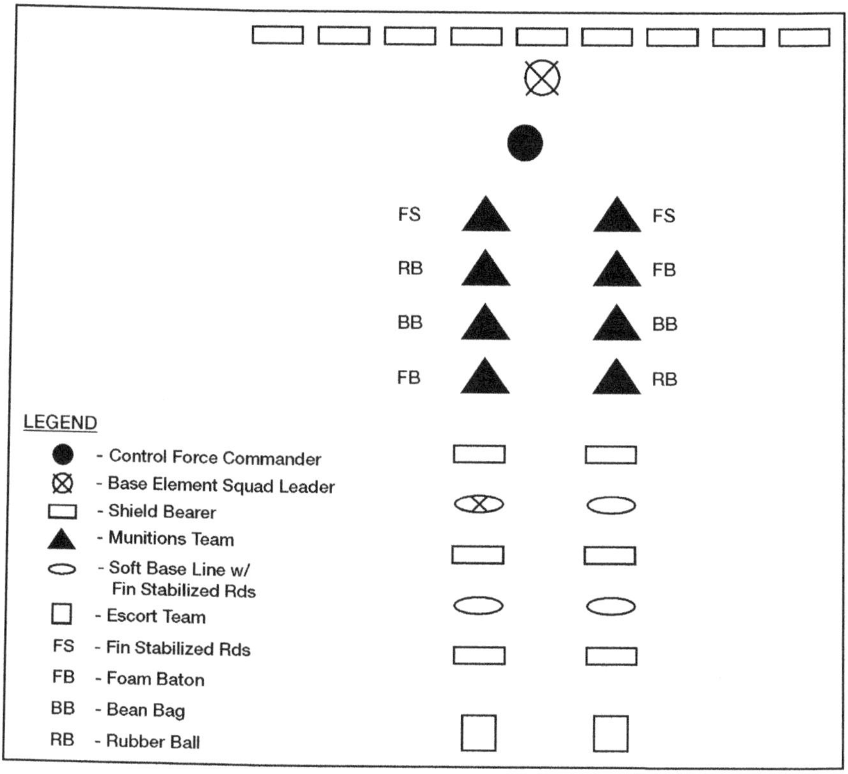

Figure V-15. Platoon in General Support (USMC)

NOTE: The control force commander moves into this formation approximately 60 to 70 meters from the crowd, as the crowd could start to engage the control force with thrown objects.

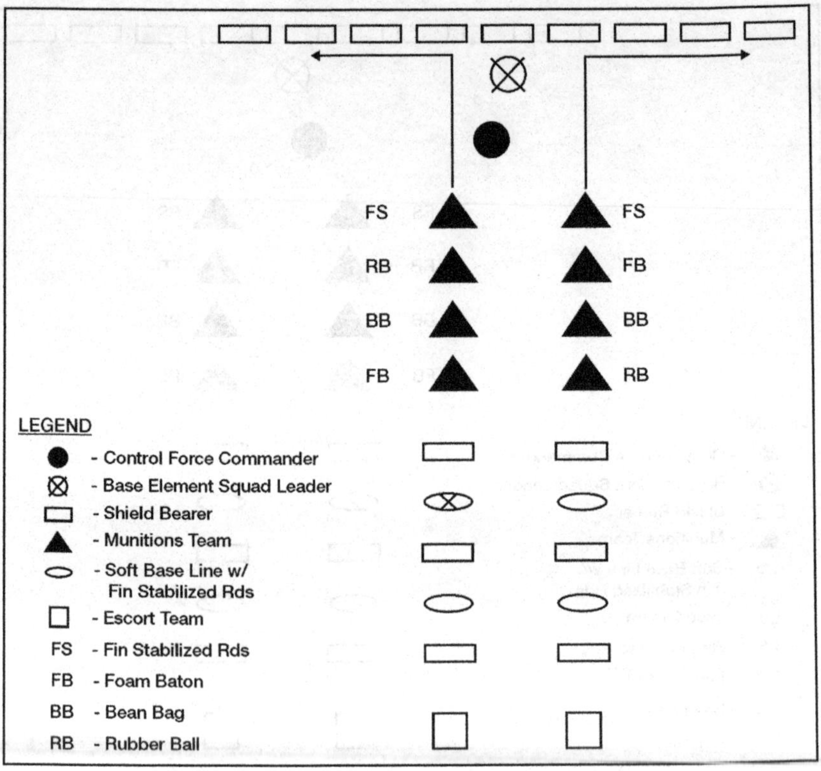

Figure V-16. Platoon in Close Support (Start) (USMC)

NOTE: The control force commander moves into this formation when he feels ready to engage the crowd with nonlethal munitions.

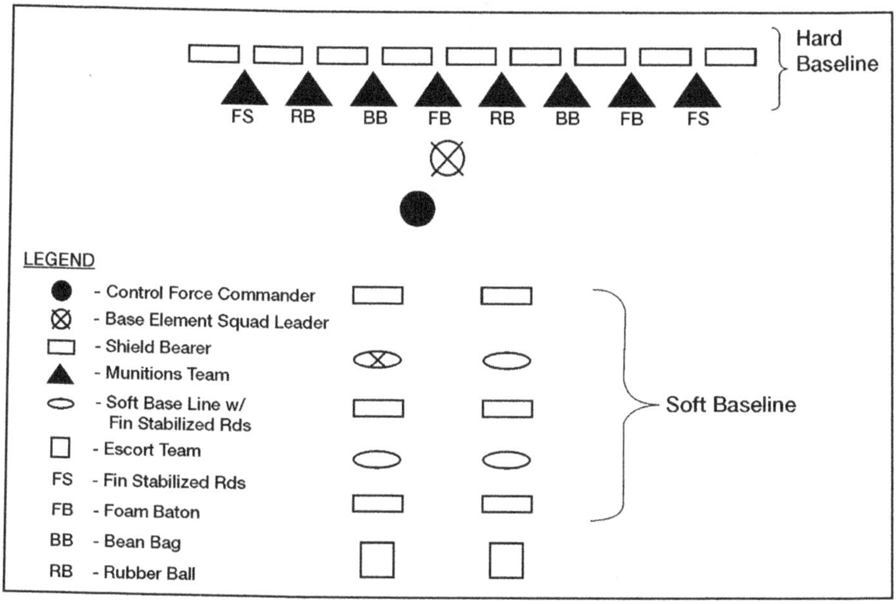

Figure V-17. Platoon in Close Support (Finish)

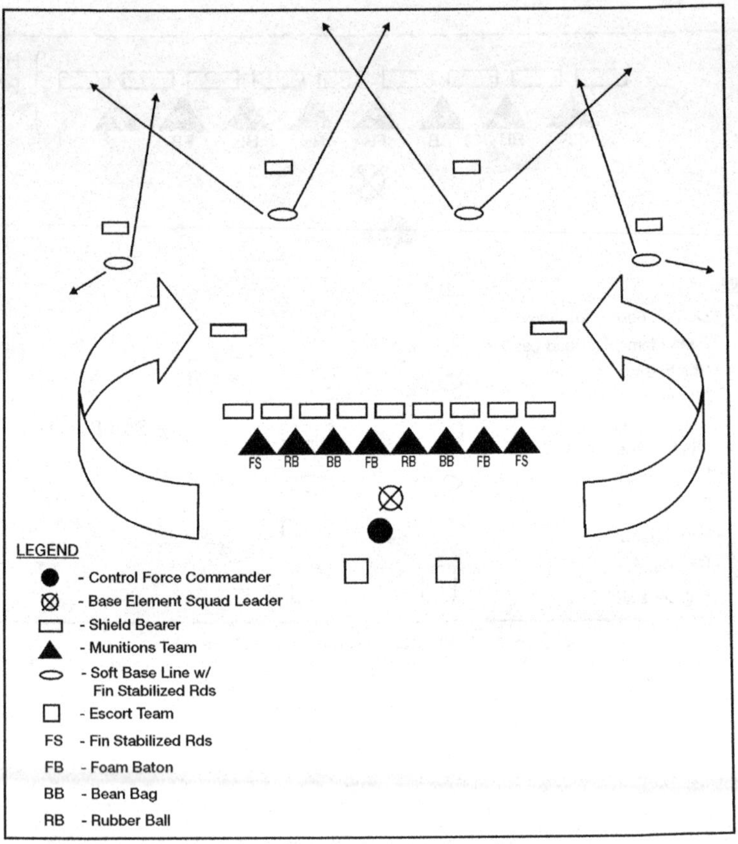

Figure V-18. Deploying Soft Baseline

***NOTE*:** The soft baseline affords the control force commander much flexibility. The primary mission of this formation is to disperse the crowd back using speed and extreme violence of action. By deploying the soft baseline, the commander is able to maneuver the hard baseline where needed (around the barrier plan or obstacles.) The soft baseline is deployed by a predetermined signal (bullhorn siren, whistle blasts, vehicle horn) and will be retrieved using the same signal. This signal conditions the crowd to expect violence of action and causes them to disperse. The soft baseline should never be deployed into a fresh crowd; the crowd needs to be distracted by nonlethal munitions. The formula found to work is distract with stingball grenades, followed by a volley of shotguns and M203 munitions, then send the soft baseline out to disperse or move the crowd back. (Sting them, shoot them, soften them up.)

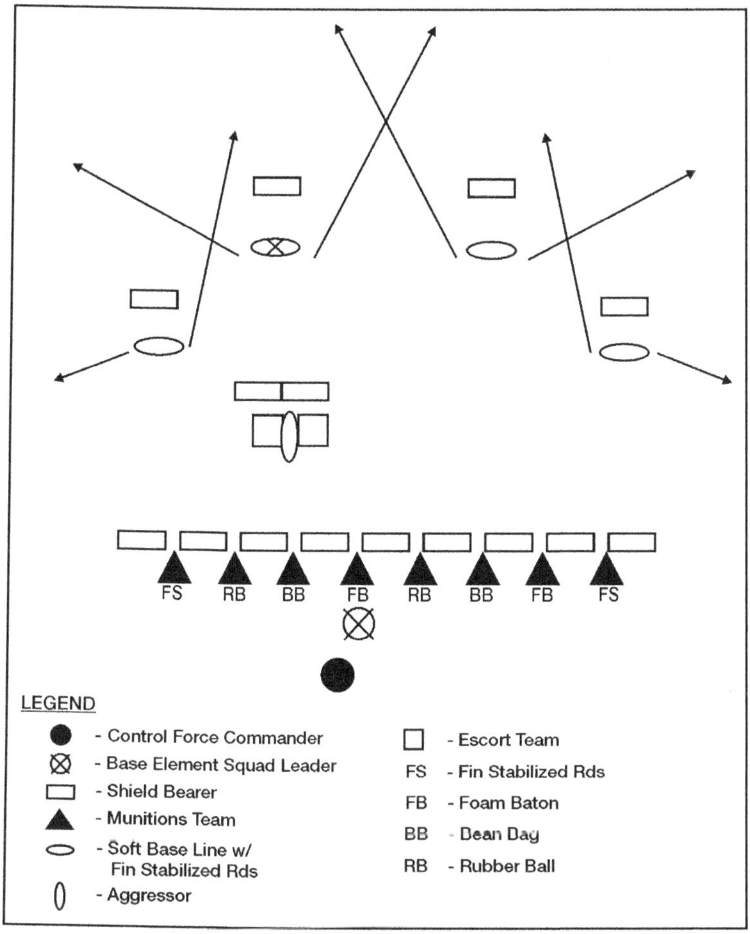

Figure V-19. Deploying Escort Team (Start)

NOTE: When the control force commander identifies the aggressor to retrieve, the command is, "Escort team left/right," then the signal. The signal is the same to deploy the soft baseline. As the soft baseline deploys, the two shields without gunners provide protection for the escort team. The shields remain with the escort team until the soft baseline is called back in, at which time the shields go back to the soft baseline.

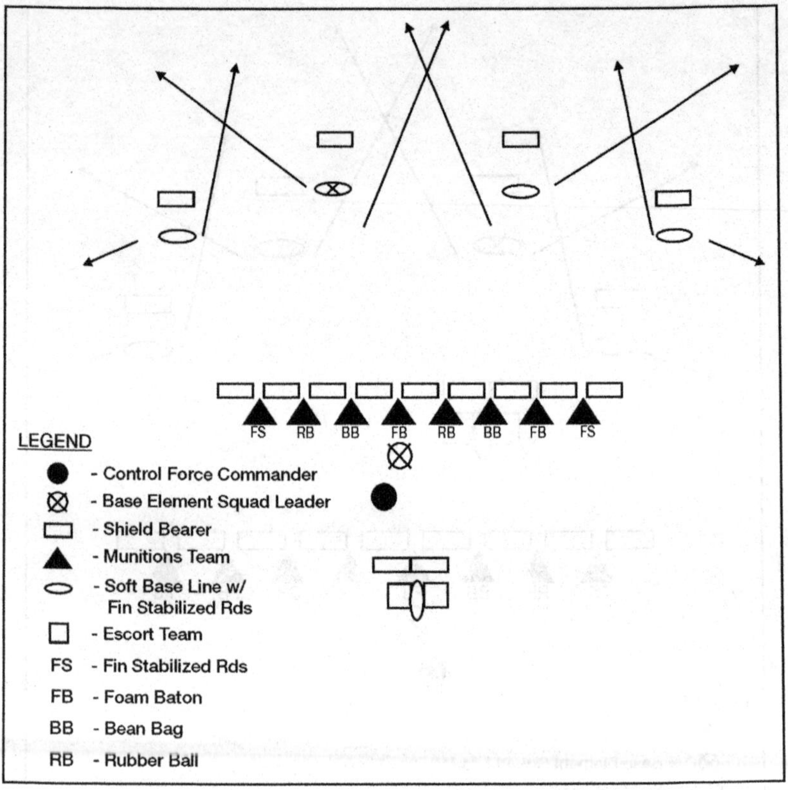

Figure V-20. Deploying Escort Team (Finish)

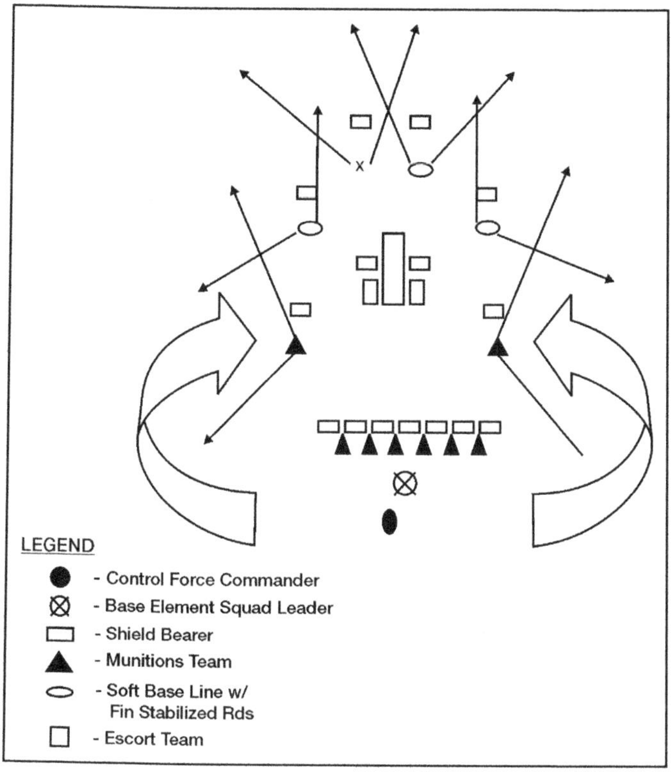

Figure V-21. Recovery of Downed Personnel or Stranded Vehicle (Big Johnson)

NOTE: In the event the crowd attacks personnel or a vehicle, the control force commander can call for a "Big Johnson." In a Big Johnson, the soft baseline deploys as usual except now the shield and shotgunners on the flank of the hard baseline also deploy creating a 190-degree perimeter around the personnel or vehicle. This maneuver must be executed quickly and with extreme violence of action because the crowd already has its hands on individuals in the force. The escort team deploys with the Big Johnson in order to escort or carry injured individuals. The shields without gunners provide protection for the escort team.

Chapter VI

LESSONS LEARNED

1. Background

Although lessons learned from employing NLW are still being studied and impacts analyzed, it is beneficial to look at the preliminary lessons gained from the NLW experience. These lessons are far from conclusive, because U.S. forces have yet to see the first large-scale employment of NLW. However, U.S. forces can gain tremendous insight from the experience of others and use what has been learned to guide future NLW employment.

2. Lessons Learned

a. Force Continuum. Using nonlethal options needs to be based on the principle of a graduated and proportional response, using the decisive force necessary to respond to the threat.

b. Attempts to legislate and regulate every aspect of NLW employment can prove disadvantageous to the on-scene commander. During the evacuation of U.N. forces in Somalia (OPERATION UNITED SHIELD, 1995), the ROE made an arbitrary and unnecessary distinction between situations justifying deadly force and those that justified only lesser means. Crowd control items were restricted to situations requiring deadly force. Many soldiers, airmen, and Marines asked, "why resort to nonlethal means at all if I have to wait until deadly force is authorized before I can shoot them with a beanbag?" Additionally, the use of sticky foam was limited to deadly force situations because of the perception that adversaries might suffocate from its application. This stipulation effectively eliminated sticky foam as a nonlethal option to the forces in Somalia.

c. With ROE that take into account the force continuum and encourage graduated response to nonlethal threats, the commander can employ NLW to enhance his flexibility, deterrence ability, and reaction to uncertain situations. To accomplish the mission with minimum loss of life and property, commanders need to have ROE that do not restrict them from properly employing NLW.

Lessons Learned from OPERATION JOINT GUARDIAN 1999 (Battalion Landing Team (BLT) 3/8 26 Marine Expeditionary Unit (MEU) Peace Operations in Kosovo:

ROE significantly impaired the employment of NLW. For example, OC was not authorized because it was considered a chemical agent. Nonlethal direct fire munitions were not authorized during the initial occupation of the province. This significantly hampered the individual Marine's ability to neutralize a threat where deadly force would not have been justified but action was required. Overly restrictive ROE thus negated the flexibility of response that NLW are supposed to provide. The intent of deploying NLW is to provide the operator the capability and flexibility to neutralize a threat without having to apply deadly force. Thus, the ROE for NLW should ideally be less restrictive, not the same or more restrictive, than those for lethal force.

It is often enticing for authors of ROE involving the use of NLW to base those ROE on the principle of minimum force necessary. This may be appropriate for base and station or homeland security ROE; however, for expeditionary operations the decision to mandate minimum force deserves critical consideration. Commanders in military operations have traditionally gained advantage over their enemy not only by the actual combat power that they bring to the fight, but also from that enemy's perceptions and beliefs about the intensity and resolve with which that combat power will be applied. When ROE are written stating that the confrontations will be addressed using the minimum force necessary, this advantage is lost. ROE developed for expeditionary nonlethal operations should be based on the concept of decisive and compelling force, rather than the minimum force. This will maintain an adversary's hesitancy to harass and endanger U.S. forces even after that adversary has tested or probed the response of those U.S. forces. Further, this will allow synergy to be gained through the complimentary employment of lethal and NLW by providing more response options to the small unit leader.

3. Employment

a. Through the use of modeling, simulation, and field exercises, new concepts for employing NLW are continually being developed. The force should train with NLW to the same degree as for lethal weapons. In this way, units become confident when using nonlethal options in chaotic and uncertain situations. The commander should push the decision to employ NLW to the lowest possible tactical level. Commanders who intend to use NLW in achieving military objectives must recognize the fluidity of the situations in which they are likely to be employed. The decisionmaking cycle is short, stressful, and the level of violence can move in both directions along the force continuum. In quick succession, a small unit leader might employ nonlethal munitions, decide to employ deadly force to an immediate specific threat that arises, then return to using nonlethal means in response to unarmed hostile elements.

b. A mixture of nonlethal "tools" has been determined to be advantageous in an operational scenario. As the situation develops, commanders can apply different nonlethal responses with respect to range/distance capability and severity of effect. For example, to disperse a crowd, area fire weapons would precede use of close-in point target nonlethal munitions.

c. An important point to remember is that NLW are just one element of a graduated response available to the force commander. Other factors make up a large contribution to the success of U.S. forces, including barrier plans and PSYOP. An example of this operation would be using barriers such as roadblocks or concertina wire, warning commands to inform a mob of potential responses based on their actions, and only then taking actions through escalated employment of NLW. These continued responses would involve smoke, RCA, and blunt munitions. A specific verbal warning should precede every escalation of nonlethal means.

d. Whenever NLW are employed, a separate lethal capability must always be in place. Troops must always be prepared to protect themselves as required by lethal means, because it is impossible to accurately predict the response of an individual or mob to NLW employment.

Lessons Learned from OPERATION ESSENTIAL HARVEST 2001 (U.S. Embassy Reinforcement operations conducted at the U.S. Embassy Skopje Macedonia by 5th Platoon, 2nd Fleet Antiterrorism Security Team [FAST] USMC):

The anticipated threats were violent demonstrations directed against U.S. facilities in the AO. In this case, USMC forces tasked to reinforce the embassy were equipped with a compliment of NLW. ROE took into account the capabilities and limitations of the NLW. The ROE and the NLW systems on the ground were coordinated and thus compatible. In addition, it was found that integrated employment of different nonlethal munitions greatly enhanced that force's ability to suppress the threats encountered (large rioting mobs). Random security patrols reinforced with military working dogs proved very effective. Reinforcement with multiple M203s provided USMC units with both a lethal and nonlethal capability in the same weapons system. The M203 was particularly useful as it gave the individual Marine the means to quickly transition between a lethal (5.56-mm) and nonlethal (40-mm nonlethal) capability. A basic understanding of the capabilities and limitations of nonlethal munitions enabled a relatively small force (40-50 Marines) to counter a sizable threat many times its size (500-1000 rioters). Although NLW were not used, the capability and preparation of the units provided the adequate deterrence that prevented the situation form escalating.

4. Publicity and Media

a. The commander can use the media to his advantage, as they will inform the local population of the military's capabilities and responses available. During OPERATION UNITED SHIELD in Somalia, the local media was informed of the Marines' nonlethal options and "new technologies" in order to discourage hostile actions. The media was encouraged to report this to the population. Intelligence had indicated the Somalis were planning to mix civilians with armed clansmen, creating hostile mobs. In fact, feuding warlords had previously successfully demonstrated this strategy on several occasions. Lack of deadly mob incidents implies Somalis were not anxious to test this new technology. Evidently, publicity had directly contributed to mission accomplishment.

b. The public dissemination of the presence of NLW capabilities must strike a balance between information sufficient to discourage hostile actions and information that enables the local population to learn too much of the capabilities of NLW. Too much information allows adversaries to develop countermeasures or avoid the NLW effects. Commanders who employ NLW systems must involve the PAO from the beginning of planning. The PAO must be familiar with the objectives of the mission, commander's intent, ROE, NLW and lethal weapons employed, and be prepared to advise the commander on the best methods of dealing with outside media. The commander should fully integrate the PA plan with PSYOP objectives and the concept of operations. Additionally, he must work the media to ensure the "sensationalism" of such a new technology does not create unrealistic expectations where he can be unfairly judged or constrained.

5. Capabilities and Impact on Mission Accomplishment

a. The commander should understand the unique capabilities of NLW and use them to accomplish the mission. He should use NLW in conjunction with the planned employment of the force and within established ROE to minimize loss of life and property while accomplishing the mission. The force requires early hands-on training to recognize the capabilities and limitations of NLW. Appendix D has sample training plans developed to meet these objectives.

b. The force commander must fully integrate planning for using nonlethal capabilities as quickly as possible in the staff planning cycle. In order to have a full range of alternatives for use against unarmed but hostile elements, the force needs to deliver nonlethal systems, complete necessary ROE and legal review, and make sufficient quantities of each NLW readily accessible to the force. The choice of nonlethal capabilities should be based on several factors and the choice can be made after addressing these questions:

(1) First, are the NLW systems available and can they be delivered on time?

(2) Second, are they compatible with, and do they complement existing weapons and training?

(3) Third, are there clear and unambiguous guidelines and ROE for employment?

(4) Finally, can the NLW be effectively employed to save lives and contribute to mission accomplishment?

c. Leaders need to anticipate local civilians developing effective countermeasures to their NLW and tactics. Therefore, prior to beginning their mission, leaders must determine and rehearse actions to take when nonlethal munitions are effectively countered.

d. Using nonlethal and lethal force are not mutually exclusive. In many instances, the commander can gain synergistic effects by using a combination of both. The following are some examples of this:

(1) Utilizing NLW to separate agitators from noncombatants with simultaneous employment of lethal force against hostiles or combatants where deadly force is justified.

(2) Utilizing NLW when making entry into structures or rooms to gain time in order to discern temper and intent of occupants prior to applying deadly force.

6. Knowing Your Adversary

a. Although commanders desire thorough intelligence of any potential adversary, such a goal may be difficult to reach in a NLW employment environment. During OPERATION UNITED SHIELD in Somalia and OPERATION UPHOLD DEMOCRACY (1995) in Haiti, a study of the local population's culture, history, and customs enabled commanders to anticipate and control many situations before they got out of hand. On the other hand, during OPERATION UPHOLD DEMOCRACY, poor intelligence and a lack of understanding between the two cultures led to the Haitians and Cubans being lodged together in Guantanamo Bay. This situation led to riots between factions who had a deep dislike for each other. By understanding a particular crowd's motivation, friendly forces may be better able to take necessary steps to prevent escalation resulting in the use of deadly force. Early training in crowd dynamics and riot control will enhance the effective use of NLW.

Lessons Learned from U.S. Peace Operations, 739th MP Battalion (Bn), January 2001

An Army MP platoon received a mission to clear a rioting crowd of 300 civilian locals from a main supply route. The platoon placed vehicle-mounted machine guns in overwatch positions to allow quick transition from NLW to conventional rounds. 12-gauge shotgun teams armed with fin stabilized rounds carried M9 pistols for lethal backup. Two volleys of shotgun and M203 nonlethal rounds successfully dispersed the crowd. The platoon had utilized INIWIC graduates to conduct preparatory training on using NLW. The NLW were extremely effective in pushing back the crowd and causing it to disperse. The 40-mm rubber ball round was moderately effective at close range, but very inaccurate at longer ranges. Shotguns were more effective than the M203 due to their high rate of fire and quick-loading capabilities.

b. Specific lessons learned during this operation are as follows:

(1) Placing interpreters is critical. The solution in this incident was to place interpreters approximately 25 meters behind formation so they could deploy forward as needed. Interpreters were often needed to communicate with crowds and with adjacent forces and coalition units that were organic to Kosovo forces (KFOR) that also responded to the incident.

(2) Leaders as well as soldiers should prepare to speak to the media. The command should have a PAO representative on hand to deal with the media.

(3) The commander should thoroughly brief the transition to lethal force from NLW.

(4) U.S. commanders in Kosovo engaged in the decisionmaking process had disparate levels of knowledge of nonlethal capabilities, constraints, and operational considerations. Therefore, leader training and orientation requires more emphasis.

(5) ROE were vague regarding the use of NLW. In many operations, NLW use is delegated to the on-site commander. Delegating to the lowest possible level is the preferred method.

(6) A unit needs a mixture of nonlethal rounds based on the size of crowd, composition of the crowd, available weapons systems, and synergistic effect of selected nonlethal combinations.

(7) When a RRF/QRF arrives on station with NLW to support a force equipped only with lethal means, the commander should address the following command and control concerns:

(a) Who makes the decision to employ NLW?

(b) Is the reinforced commander familiar with NLW?

(c) Who controls the reinforced unit?

(d) How does this relationship change as the situation develops?

7. Summary

a. The wide range of military operations in which U.S. forces are involved has demanded the introduction of NLW into the arsenal of weaponry available to commanders. NLW provide a graduated and proportional response to threats posed by a variety of adversaries. The ROE must be consistent with the rapidly changing situation of today's chaotic and uncertain military operations. This reality demands a high level of training in properly employing NLW.

b. Effective use of NLW involves a synergy between PSYOP activities, intelligence gathering, and professional media/PA plan. NLW effectiveness is enhanced through integration of these activities to reduce the chance adversaries will be able to develop effective countermeasures. NLW, when combined with lethal weapons, provide the commander with a force response across a wide range of military operations.

Appendix A

DOD Nonlethal Weapons and Equipment Item Description

The following is a cumulative list of NLW and equipment currently fielded within the DOD. Each service has various components of the following list in its inventory. A list of the service's items is located in Appendix E.

Figure A-1. Riot Face Shield

1. Riot Face Shield: The riot face shield is a non-ballistic face shield designed to protect the service member's face from objects thrown or attacks from non-ballistic weapons. (Figure A-1)

No picture available.

2. Ballistic Face Shield: The ballistic face shield provides the service member with face protection from a wide range of threats (i.e. debris, liquids, handthrown objects). The shield is manufactured from acrylic and bullet resistant materials to provide Level IIIA protection (9-mm and .44 magnum pistol rounds), weighs 3.4 pounds and is compatible with the PASGT helmet and M-17/M-40 protective masks.

Figure A-2. Riot Body Shield

3. Riot Body Shield: The riot body shield is a non-ballistic body shield designed to protect the service member's body from objects thrown or attacks from non-ballistic weapons. (Figure A-2)

No picture available.

4. Ballistic Body Shield with Light Kit: The ballistic shield provides the service member with torso protection from a wide range of threats (i.e. debris, liquids, hand thrown objects). The shield is manufactured from Spectra Shield resistant material painted anti-reflective flat black, and is capable of providing Level IIIA protection (9-mm and .44 magnum pistol rounds). The shield weighs 18 pounds and is 20 x 36 inches in size with a view port of 4 x 16 inches.

 **Figure A-3. Riot Shin Guard**	5. Riot Shin Guard: The riot shin guard is a non-ballistic shin guard designed to protect the service member's shins from objects thrown or attacks from non-ballistic weapons. (Figure A-3)
No picture available.	6. Ballistic Shin Guards: Ballistic shin guards provide the service member with leg protection from a wide range of threats (i.e. debris, liquids, hand thrown objects). The guards are manufactured from Kevlar KM2 and provide Level IIIA protection (9-mm and .44 magnum pistol rounds). They are available in three sizes and weigh between 7 to 10 pounds.
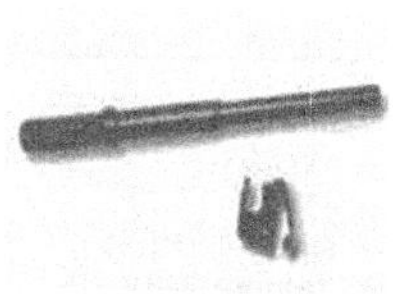 **Figure A-4. Expandable Riot Baton (with holster)**	7. Expandable Riot Baton (with holster): The expandable riot baton provides the service member a nonlethal means of crowd control and self-defense. The baton comes with a mounting device that attaches the baton to a belt. The baton is 24 to 36 inches in length. (Figure A-4)
No picture available.	8. Wooden Baton: The wooden baton provides self-defense for the service member and the option between physical force and lethal force. The baton comes with a mounting device that attaches the baton to a belt. The baton is 36 inches in length.
No picture available.	9. Practice Riot Batons: The practice riot batons provide the service member with a safe dynamic training environment for properly using a riot baton. The batons are available in a variety of lengths.
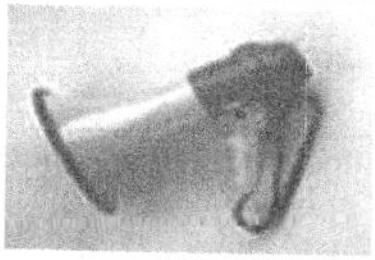 **Figure A-5. Portable Bullhorn**	10. Portable Bullhorn: The portable bullhorn provides the service member the capability to control his forces by projecting his voice above the noise and commotion created by crowds and mobs. (Figure A-5)
No picture available	11. Ground-Mounted Public Address System: The ground-mounted public address system is a communication enhancement device for conducting crowd control tactics. The system facilitates communication with the crowd in the absence of PSYOP support and assists with communication of commands to troops engaged in crowd-over-crowd noise.
No picture available.	12. M7 Individual Voice Amplification System: The M7 individual voice amplification system is a communication enhancement device for conducting crowd control tactics while wearing the M40 protective mask and using riot control agents. The M7 is fitted to the M40 protective mask to facilitate oral communication and increase the user's ability to communicate using radios and other devices.

Figure A-6. Disposable Handcuff

13. Disposable Handcuff: The disposable handcuff provides service members with a means of restraining and controlling suspects, prisoners, and other individuals encountered during crowd control. The disposable handcuff is designed for one-time use. (Figure A-6)

Figure A-7. Training Disposable Handcuff

14. Training Disposable Handcuff: The training disposable handcuff provides service members with a means for training in handcuffing procedures prior to actual use. The training disposable handcuff is red and reusable. (Figure A-7)

Figure A-8. MK-4 Pepper Spray Pouch

15. MK-4 Pepper Spray Pouch: The MK-4 pepper spray pouch is designed to carry the MK-4 pepper spray canister. (Figure A-8)

Figure A-9. MK-4 Individual Live OC Pepper Spray

16. MK-4 Individual Live OC Pepper Spray: The MK-4 individual pepper spray has a range of up to 12 feet in a target-specific stream. It contains 30 to 35 half-second bursts. (Figure A-9)

No picture available.

17. M-36 Individual Riot Control Disperser: M-36 Individual Riot Control Disperser has a range of up to 12 feet in a target-specific stream. It contains 30 to 35 half-second bursts. It is loaded with either CN or CS agent.

Figure A-10. MK-9 Pepper Spray Pouch

18. MK-9 Pepper Spray Pouch: The MK-9 pepper spray pouch is designed to carry the MK-9 pepper spray canister. (Figure A-10)

Figure A-11. MK-9 Squad Live OC Pepper Spray

19. MK-9 Squad Live OC Pepper Spray: The MK-9 squad pepper spray has a range of up to 15 feet in a target-specific fog. It contains 8 to10 half-second high volume bursts. (Figure A-11)

Figure A-12. MK-46 Platoon Live (Horizontal) OC Pepper Spray

20. MK-46 Platoon Live (Horizontal) OC Pepper Spray: The MK-46 platoon pepper spray has a range of up to 25 to 30 feet in a target-specific dispersed spray pattern. It contains up to 12 one-second high volume bursts. (Figure A-12)

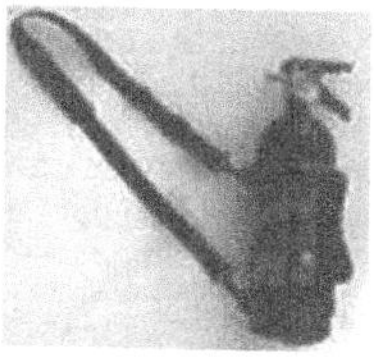

Figure A-13. MK-46 Sling

21. MK-46 Sling: The MK-46 sling gives a service member the ability to sling the MK-46 canister on the shoulder. (Figure A-13)

Figure A-14. MK-46 Hardware Kit

22. MK-46 Hardware Kit: The MK-46 hardware kit consists of a transfer tube, funnel, and transfer bottle. A compressed nitrogen source is also required to utilize the kit. (Figure A-14)

Figure A-15. MK-46 Refill Live OC Pepper Spray

23. MK-46 Refill Live OC Pepper Spray: The MK-46 refill solution comes in a one-gallon container. (Figure A-15)

Figure A-16. MK-4 Individual Inert OC Pepper Spray

24. MK-4 Individual Inert OC Pepper Spray: The MK-4 inert individual pepper spray has a range of up to 12 feet in a target-specific stream. It contains 30 to 35 half-second bursts. It provides the service member with an inert canister for familiarization and training purposes. The inert device is loaded with a non-irritant, non-toxic formulation payload. The device may also contain a non-toxic, water-soluble marker to allow immediate feedback on content delivery to target. The MK-4 device is marked "Inert." (Figure A-16)

Figure A-17. MK-9 Squad Inert OC Pepper Spray

25. MK-9 Squad Inert OC Pepper Spray: The MK-9 squad inert pepper spray has a range of up to 15 feet in a target-specific fog. It contains 8 to 10 half-second high volume bursts. It provides the service member with an inert canister for familiarization and training purposes. The inert device is loaded with a non-irritant, non-toxic formulation payload. The device may also contain a non-toxic, water-soluble marker to allow immediate feedback on content delivery to target. The inert MK-9 device is prominently marked "Inert" with a yellow sticker attached. (Figure A-17)

Figure A-18. MK-46 Platoon Inert OC Pepper Spray

26. MK-46 Platoon Inert OC Pepper Spray: The MK-46 platoon inert pepper spray has a range of up to 25 to 30 feet in a target-specific dispersed spray pattern. It contains up to 12 one-second high volume bursts. It provides the service member with an inert canister for familiarization and training purposes. The inert device is loaded win a non-irritant, non-toxic formulation payload. The device may also contain a non-toxic, water-soluble marker to allow immediate feedback on content delivery to target. The inert MK-46 device is prominently marked "Inert" with a yellow sticker attached. (Figure A-18)

Figure A-19. MK-46 Refill Inert OC Pepper Spray

27. MK-46 Refill Inert OC Pepper Spray: The MK-46 refill solution comes in a one-gallon container. (Figure A-19)

Figure A-20. FIST Training Suit

28. FIST Training Suit: The FIST training suit is made of closed-cell shock-absorbent foam that allows service members to hone their riot baton skills through realistic hand-to-hand engagement training. The FIST suit can absorb blows inflicted by the current expandable and wood riot baton. It provides protection for the head, face, hands, and legs, without significant degradation of the wearer's mobility. The FIST suit consists of a helmet, chest and back protector, bicep and forearm protectors, groin and buttock protectors, thigh and shin protectors, gloves, and carry bag. (Figure A-20)

NOTE: FIST is the current manufacturer of this training suit).

Figure A-21. Riot Training Strike Bag

29. Riot Training Strike Bag: The riot training strike bag gives the service member a strike bag for training and proficiency in open-hand control techniques. (Figure A-21)

Figure A-22. Caltrop

30. Caltrop: The caltrop is a field expedient area denial system. Securing approximately 50 caltrops to 550 military cord best employs it. This allows a service member to quickly toss/recover the system, using the caltrops to deny vehicle/pedestrian movement in a designated area. The four prongs are approximately two inches tall and will puncture tires or boots. (Figure A-22)

Figure A-23. Road Side Spike Strip

31. Road Side Spike Strip: The spike strip is designed to puncture a pneumatic tire as the vehicle makes contact with the spike strip. The spikes pivot into the tire and the tip-guards are pushed down. The spikes are imbedded deeply into the tire. The spikes are pulled out of the unit and remain in the tire. On the second revolution they will be pushed directly into the tire, allowing air to escape slowly without causing a blowout or sudden loss of control. The average tire will pick up 4 to 6 spikes with the deflation time of about 12 to 20 seconds. It allows a service member to quickly toss/recover the system, impeding vehicle movement in a designated area. It allows a force to provide additional physical security in areas where the caltrop cannot be tactically employed. (Figure A-23)

Figure A-24. High-Intensity Search Light

32. The high-intensity searchlight is a rugged, lightweight, hand-held, high-intensity spotlight for use during low-light conditions. Two different high-intensity lights were fielded. The Maxabeam light (shown on top) was fielded in 1997. It was replaced by the Nighthunter light (shown underneath) in 2000. (Figure A-24)

Figure A-25. Individual Light Holster (6P)

33. Individual Light Holster (6P): The 6P individual light holster is designed to hold the 6P individual light. (Figure A-25)

Figure A-26. 6P Individual Light

34. 6P Individual Light: The 6P individual light is issued to service members within the platoon. (Figure A-26)

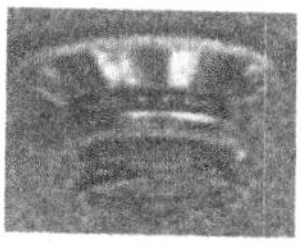

Figure A-27. 6P Individual Light Accessory Kit

35. 6P Individual Light Accessory Kit: The 6P individual light accessory kit includes a spare light bulb installation kit. (Figure A-27)

Figure A-28. 3V Lithium Batteries

36. 3V Lithium Batteries: Two 3V Lithium Batteries are required to operate the 6P individual light. (Figure A-28)

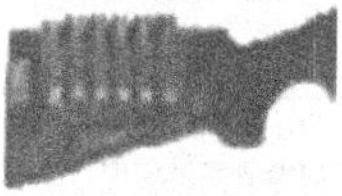

Figure A-29. 12-Gauge Buttstock Cuff

37. 12-Gauge Buttstock Cuff: The buttstock cuff is designed to fasten to the buttstock of most shotguns to hold seven 12-gauge cartridges for quick loading of the shotgun. (Figure A-29)

Figure A-30. 12-Gauge Utility Pouch

38. 12-Gauge Utility Pouch: The 12-gauge utility pouch is designed to hold twenty-five 12-gauge munitions and attaches to the service member's combat gear. (Figure A-30)

Figure A-31. 40-mm Carrying Pouch

39. 40-mm Carrying Pouch: The 40-mm carrying pouch is partitioned in a manner to provide the service member with a means of carrying not less than six 40-mm nonlethal rounds in separate compartments. It attaches to the service member's combat gear. (Figure A-31)

Figure A-32. Diversionary/Rubber Ball Grenade Pouch

40. Diversionary/Rubber Ball Grenade Pouch: The diversionary/rubber ball grenade pouch is an ammunition carrying pouch partitioned in a manner to provide the service member with a means of holding up to six rubber ball grenades or six flash bang grenades in separate compartments. (Figure A-32)

Figure A-33. 12-Gauge Mossberg 500A2/590 Shotgun

41. 12-Gauge Mossberg 500A2/590 Shotgun: The 12-gauge Mossberg 500A2 or 590 shotgun are the only 12-gauge shotguns currently authorized for firing 12-gauge nonlethal projectile munitions. The launching cup for the rubber ball grenade will attach to the Mossberg 500A2 Shotgun. The launch cup will not attach to the Mossberg 590 shotgun. Until the Joint Service Combat shotgun can effectively fire nonlethal 12-gauge rounds, the 12-gauge Mossberg 500A2 or 590 shotgun configuration meets the requirement of the NLW force. (Figure A-33)

Figure A-34. 12-Gauge M870 (modified) Version 2 (USCG)

42. 12-Gauge M870 (modified) Version 2 (USCG): The modified Remington 870 shotgun has a fixed synthetic stock and pistol grip. The pistol grip gives the shooter more control of the weapon and helps dissipate recoil from the shoulder. The synthetic materials used are more durable and lighter weight than the wood stocks used on the Standard M870. The barrel is fitted with a rifled choke tube for firing slugs (solid projectiles) to increase accuracy. It is also equipped with an Advanced Combat Optical Gunsight that was developed for use in "both-eyes-open" close quarters battle. The advantage is rapid target acquisition without the need to align front and rear sights. Batteries do not power the illumination sight; it contains tritium gas.

In the Coast Guard, a yellow "stock sock" will be used whenever the shotgun is loaded with nonlethal munitions. This removable nylon sleeve is used to identify the weapon as a nonlethal delivery system. (Figure A-34)

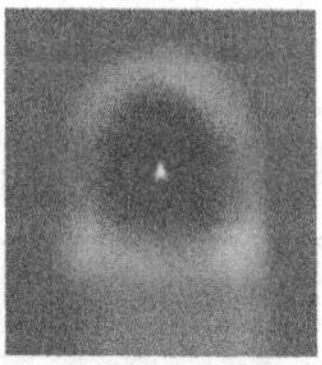

Figure A-35. Advanced Combat Optical Gunsight (ACOG) Reflex Sight (USCG)

43. Advanced Combat Optical Gunsight (ACOG) Reflex Sight (USCG): Modified shotguns are equipped with this site. (Figure A-34)

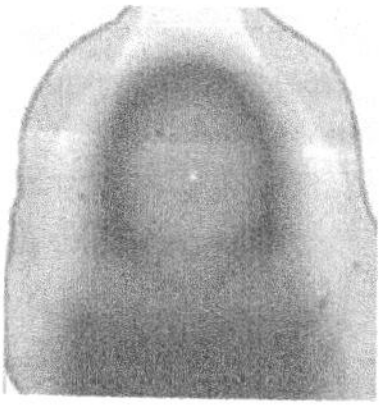

Figure A-36. ACOG Reflex Sight (USCG)

44. ACOG Reflex Sight (USCG): The dual-illuminated, Trijicon® Reflex sight was developed for the military for use in both-eyes-open close quarters battle (CQB). The advantage is quick target acquisition thus eliminating the need to align front and rear sights. Batteries do not power the illumination sight; it contains tritium gas. (Figure A-35)

Equipment Currently Being Fielded

1. 40-mm Crowd Dispersal Cartridge (CDC):

This direct fire, low-hazard, non-shrapnel-producing, multiple ball projectile provides friendly forces the capability to conduct crowd control. The CDC uses nonlethal, blunt trauma to stop, confuse, disorient, and/or temporarily incapacitate personnel and increase standoff distance between friendly forces and the hostile crowd. Primarily designed as an area effect munition, the CDC can be used on individuals. The M203 40-mm grenade launcher fires the CDC and contains 48, .48 caliber rubber balls. Effective range is 10 to 30 meters. (Figures A-37 and A-38)

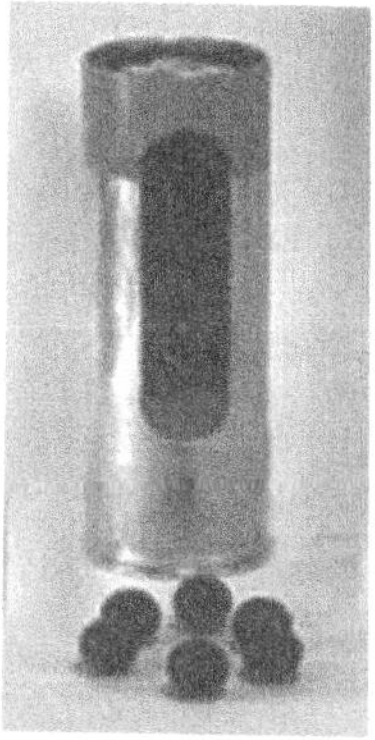

Figure A-37. Crowd Dispersal Cartridge

Figure A-38. Soldier Employing CDC with M203

2. Portable Vehicle Arresting Barrier (PVAB):

PVAB is a lightweight, portable, manually emplaced and recoverable barrier entrapment device to create, on command, a barrier that stops vehicles driven by known or suspected hostile forces with minimal damage to the vehicle or its occupants. It is used at dismount points, checkpoints, and roadblocks to prevent unauthorized access by wheeled vehicles into or out of areas (e.g. airfields, ports, command posts [CP]). Two individuals can easily emplace it, in approximately four hours. In stand-by mode, PVAB folds down into something similar to a speed bump (vehicles pass freely). In capture mode, the net springs up on two hydraulic pistons, to maximum height in less than a second. PVAB can be set up to span 12 to 24 feet and can stop a light truck (up to 7,500 pounds) at speeds up to 45 miles per hour, with a stopping distance of less than 200 feet PVAB can be containerized: 53.75 cubic feet (7 boxes, 4 large-3 small)—1,044 pounds on approximately one pallet. (Figures A-39 and A-40)

Figure A-39. PVAB in Action (Far View)

Figure A-40. PVAB in Action Near View

3. Modified Crowd Control Munition—Ground Emplaced (MCCM-GE):

The MCCM-GE (Figure A-41) is a nonlethal munition, housed in a casing similar to the current claymore mine, which will stop, confuse, disorient and/or temporarily incapacitate area targets/personnel at close range. The force can use it to provide crowd control by dispersing hostile crowds and force protection by integrating it into a layered perimeter defense. It delivers 600, .32 caliber rubber balls as well as a flash bang effect. The MCCM-GE gives the commander or small unit leader greater flexibility to apply nonlethal force as a first line of defense against hostile noncombatants. Effective range is 5 to 15 meters. (Figure A-42)

Figure A-41. Modified Crowd Control Munition-Ground Emplaced (MCCM-GE)

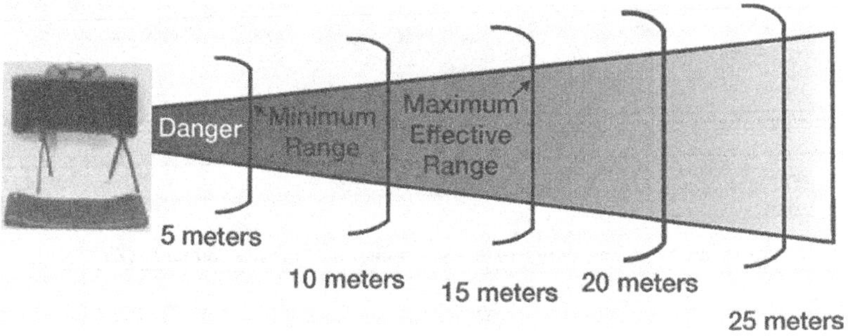

Figure A-42. MCCM-GE Range Chart

4. 66-mm Vehicle Mounted Nonlethal Grenade Launcher (VMNLGL):

The VMNLGL is a nonlethal grenade-launching device mounted on high mobility multipurpose wheeled vehicle (HMMWV). It is an indirect fire-support system that can deliver a 66-mm distraction device (flash-bang), CS, or blunt trauma (sting ball) payload. The distraction grenade uses a pyrotechnic charge for crowd control through audio and visual stimuli. The blunt trauma grenade uses a pyrotechnic charge to discharge 140, .32 caliber (cal.), rubber balls to achieve crowd control through audio, visual, and physical stimuli. The force uses VMNLGL to enhance the ability of friendly forces to conduct force protection without direct contact with hostile or threatening crowds. It enforces standoff distances and deters potential threats. It can disperse or subdue crowds, keep personnel from surrounding or climbing on patrol vehicles, slow or stop threatening personnel or vehicles by confusing, disorienting, and/or temporarily incapacitating individuals for capture. The grenade is fired from the grenade launchers traditionally found on tracked vehicles. Vehicle-launched grenades have three submunitions designed to ground burst. Effective range is 50 to 100 meters. An adjustable bracket allows engagement at ranges of 50 meters, 75 meters, and 100 meters. (Figure A-44)

Figure A-43. 66-mm Vehicle Mounted Nonlethal Grenade Launcher (VMNLGL)

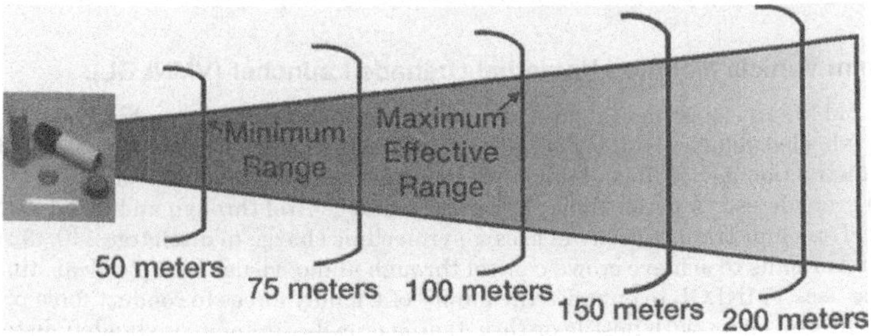

Figure A-44. VMNLGL Range Chart

5. Running Gear Entanglement System (RGES)

The Running Gear Entanglement System (RGES) is series of short, separate, looped polyethylene lines that are attached to a single, long polyethylene base line. This entanglement net fouls the propellers of planing hull small boats. The capability to manually deploy the net from a soft-sided bag has been fielded for USCG use in limited numbers. The USCG is in the process of incorporating three other variants of RGES into its inventory of nonlethal capabilities. These include aerial-delivered RGES, surface launched RGES to deploy from vessels, and static barrier RGES. The static barrier RGES has broad application into support homeland security, physical security and force protection missions. The surface launched RGES has direct application in the USCG's law enforcement mission wherever the requirement exists to stop a non-compliant vessel. Both of these systems are expected to be fielded in the fall/winter of CY02. The air launched version is currently still in the concept development phase and is not yet fully developed. (Figure A-45)

Figure A-45. Running Gear Entanglement System (RGES)

a. RGES Surface Delivery System: The USCG can use stern-mounted, manually deployed RGES from a cutter's small boat to stop slow-moving migrant-type craft. (Figure A-46)

(1) Uses COTS compressed air cylinder launcher.

(2) Removable, fix-mounted launcher on side of craft.

(3) Launch projects device ahead of target.

Figure A-46. RGES Surface Delivery System

b. RGES Barrier

(1) Provides static barrier around ship or other waterside asset.

(2) Provides a visible demarcation line, establishes zones, deters intrusions. (Figure A-47)

Figure A-47. RGES Barrier Employment

c. RGES Aerial Delivery System (ADS): RGES is aerial deployed from the ADS ahead of target aboard a H60 helicopter (Figure A-48). Device should float in the water ahead of target ensuring target runs over device. Device fouls propeller or rudder until target stops and awaits boarding.

(1) Uses COTS compressed air cylinder launcher

(2) Removable, fix-mounted launcher on helicopter

(3) Launch projects device ahead of target. (Figures A-49)

Figure A-48. RGES Aerial Delivery System (ADS)

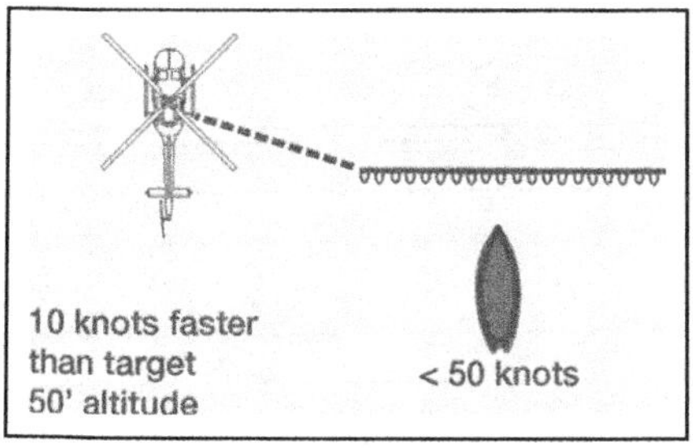

Figure A-49. RGES Aerial Delivery System (ADS)

6. Warp Cutter

a. Warp cutting is a nonlethal use of force, which has been safely and effectively employed by several countries to enforce fisheries laws in their respective exclusive economic zones (EEZ). The warp cutter is a device that closely resembles a grappling hook or small anchor. It has hardened steel blades attached to the crux of the grapnels. It is towed behind an enforcement vessel and across the warps of a non-compliant fishing vessel. As the cutting device engages the warps, it slices through them, separating the net from the fishing vessel.

b. The USCG uses the warp cutter to compel foreign vessels suspected of illegally fishing in the U.S. EEZ to comply with lawful orders to stop and be boarded. The USCG may cut the trawl warps of non-compliant vessels as a means of compelling them to stop for a boarding.

Figure A-50. Warp Cutter

Appendix B

Summary: Specialty Impact Munitions

12-Gauge Launch Cartridge

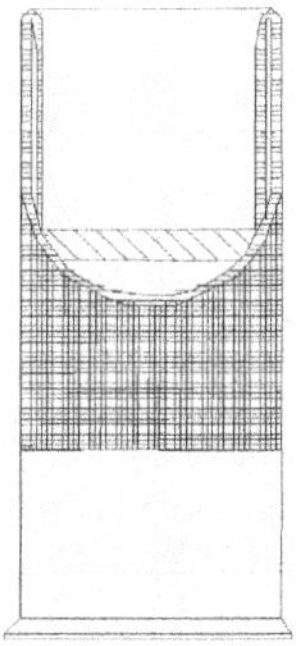

Using Services: USN, USMC
Status: Fielded 1998
DODIC: AA30
NSN: 1305-01-464-8389
Acquisition Activity: MARCORSYCOM—NSWC Crane, IN
Further Information:

Commercial: MARCORSYSCOM (703) 784-9395
NSWC Crane (812) 854-5801

Color ....Clear case
Diameter 12-gauge
Projectile ....N/A
Length ....2.00 inches (51 mm)
Total Weight..........................13.6 grams
Propellant Weight.................46.3 grains (black powder)
Muzzle Velocity.....................100 feet/second (30.5 m/sec)—rubber ball grenade
Acceptance Accuracy..........200 feet (61 m) + 25 feet (7.6 m) at 30-degree launch angle using rubber ball grenade and 12-gauge launch cup

Projectile Specification: N/A

- Use 12-gauge launch round in combination with rubber ball grenade and 12-gauge launch cup to hurl a grenade a distance of 200 feet (61 m) + 25 feet (7.6 m) when launch angle is approximately 30 degrees.
- Chamber launch cartridge only after installing grenade in launch cup and pulling safety pin from grenade.

Figure B-1. 12-Gauge Launch Cartridge

12-Gauge Bean Bag Cartridge

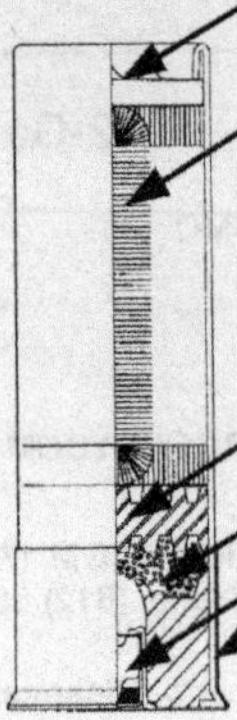

Using Services: USN, USMC
DODIC: AA29
NSN: 1305-01-454-0191
Status: Fielded 1998
Acquisition Activity: MARCORSYCOM—NSWC Crane, IN
Further Information:
Commercial: MARCORSYSCOM (703) 784-9395
NSWC Crane (812) 854-5801

Color......................................**Clear or red case**
Diameter................................**12 gauge**
Projectile...............................**1 bean bag**
Length....................................**2.45 inches (62 mm)**
Total Weight..........................**49.9 grams**
Propellant Weight.................**3.4 grains**
Muzzle Velocity.....................**300 + 25 feet/second (91.4 + 7.62 m/sec)**
Acceptance Accuracy..........**2 feet (.6 m) X 6 feet (1.8 m) at 100 feet (30.3 m) with Mossberg 500A2**

Shotgun Projectile Specification:
Shape..............................**Square bag or tube sock**
Material............................**Nylon or cotton-polyester**
Filler.................................**#9 lead shot**
Filler weight.....................**44.4 grams (maximum)**

- **Fatal injuries are possible at employment distances of less than 20 feet (6 m)**
- **At distances of 20 feet (6 m) to 40 feet (12 m), target area should be center mass**
- **Training needs to reinforce that head shots are NOT ACCEPTABLE**
- **Do not skip fire this round! Round becomes unpredictable upon striking the ground**
- **Round is most effective against individually selected targets (point round)**
- **End wad has low light marker**

Figure B-2. 12-Gauge Bean Bag Cartridge

12-Gauge Rubber Fin Stabilized Cartridge

Using Services: USN, USMC, USCG

Status: Fielded 1998

DODIC: AA31

NSN: 1305-01-454-0189

Acquisition Activity: MARCORSYCOM—NSWC Crane, IN

Further Information:

Commercial: MARCORSYSCOM (703) 784-9395
NSWC Crane (812) 854-5801

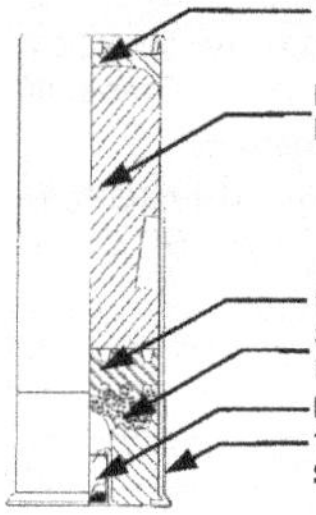

Color .... Clear or red case
Diameter .... 12 gauge
Projectile .... 1 rubber projectile
Length .... 2.45 inches (62 mm)
Total Weigh .... 9.1 grams
Propellant Weight................. 4.8 grains
Muzzle Velocity..................... 530 + 65 feet/second
(161.5 + 19.8 m/sec)
Acceptance Accuracy.......... 2 feet (.6 m) X 6 feet (1.8 m) at 100 feet (30.3 m) with Mossberg 500 A2 shotgun
Projectile Specification: ...... Caliber 12 gauge
Material .... Rubber
Hardness .... Durometer 70-80 Shore "A" scale
Weight .... 7.8 grams
Temperature Limits (demonstrated):
Upper limit...125° F
Lower limit...0° F

Shipping Data: Hazard classification/division and storage compatibility group...1.4S
DOT Class .... ORM-D
DOT Markings Cartridge...... Small Arms
UNO serial number............... 0012

- Fatal injuries are possible at employment distances of less than 20 feet (6 m); USCG: 15 yards
- At distances of 20 feet (6 m) to 40 feet (12 m), target area should be center mass; USCG: 15 to 30 yards
- Training needs to reinforce that head shots are NOT ACCEPTABLE
- Do not skip fire this round! Round becomes unpredictable upon striking the ground
- Round is most effective against individually selected targets (point round)
- End wad has low light marking

Figure B-3. 12-Gauge Rubber Fin Stabilized Cartridge

40-mm Foam Rubber Baton Cartridge

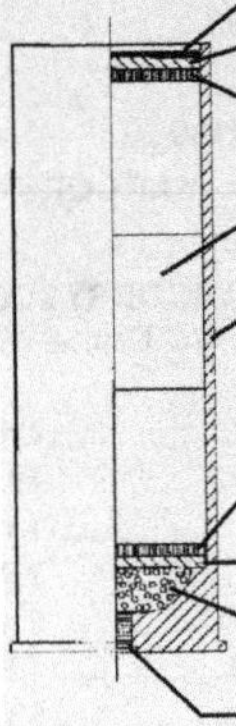

Using Services: USN, USMC

Status: Fielded 1998

DODIC: BA07

NSN: 1310-01-453-9168

Acquisition Activity: MARCORSYSCOM—NSWC Crane, IN

Further Information:

Commercial: MARCORSYSCOM (703) 784-9395
NSWC Crane (812) 854-5801

Color .... **Aluminum (case)**
Diameter .... **40 millimeter**
Projectile .... **3 foam batons**
Length .... **4.78 inches (121.4 mm)**
Total Weight .... **201 grams**
Propellant Weight .... **13.0 grains**
Muzzle Velocity .... **325 + 35 feet/second (99 + 10.6 m/sec)**
Acceptance Accuracy .... **6 feet (1.8 m) X 3 feet (.9 m) at 50 feet (15.2 m) with M203**

Projectile Specifications:

Number **3**
Material .... **Foam rubber**
Hardness .... **Durometer 25 + 10 on Shore "A" scale**
Weight .... **250 to 280 grains**

- **Fatal injuries are possible at employment distances of less than 20 feet (6 m)**
- **At distances of 20 feet (6 m) to 40 feet (12 m), target area should be center mass**
- **Training needs to reinforce that head shots are NOT ACCEPTABLE**
- **Round is most effective against point targets (individual personnel)**
- **End wad has low light marker**

Figure B-4. 40-mm Foam Rubber Baton Cartridge

40-mm Rubber Ball Cartridge

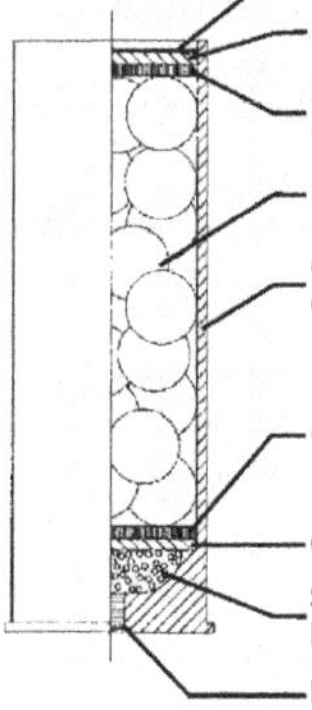

Using Services: USN, USMC

Status: Fielded 1998

DODIC: BA08

NSN: 1310-01-453-9154

Acquisition Activity: MARCORSYSCOM—NSWC Crane, IN

Further Information:

Commercial: MARCORSYSCOM (703) 784-9395

Color .... Aluminum (case)
Diameter .... 40 millimeter
Projectile .... 24 rubber balls
Length .... 4.78 inches (121.4 mm)
Total Weight .... 201 grams
Propellant Weight .... 7.3 grains
Muzzle Velocity .... 325 + 35 feet/second (99 + 10.6 m/sec)
Acceptance Accuracy .... 6 feet (1.8 m) x 9 feet (2.7 m) at 50 feet (15.2 m) with M203NSWC Crane (812) 854-5801

Projectile Specifications:

Caliber 60
Number .... 22 (minimum)
Material .... Rubber
Hardness .... Durometer 60 + 10 on Shore "A" scale
Weight .... 2.2 grams each

- Fatal injuries are possible at employment distances of less than 20 feet (6 m)
- At distances of 20 feet (6 m) to 40 feet (12 m), target area should be center mass
- Training needs to reinforce that head shots are NOT ACCEPTABLE
- Round is most effective against area targets (multiple personnel)
- End wad has low light marker

Figure B-5. 40-mm Rubber Ball Cartridge

Rubber Ball Grenade

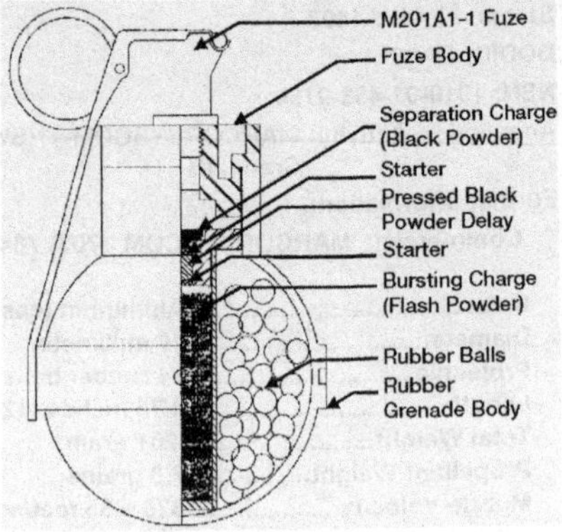

Using Services: USN, USMC
Status: Fielded 1998
DODIC: GG04
NSN: 1310-01-454-0132
Acquisition Activity: MARCORSYSCOM—NSWC Crane, IN
Further Information:
Commercial: MARCORSYSCOM (703) 784-9395
NSWC Crane (812) 854-5801

Color	Black with white lettering
Diameter	3.12 inches (79 mm)
Projectiles	100 rubber balls (minimum)
Height	5.25 inches (133.4 mm)
Total Weight	290 grams
Explosive Weight	59 grams
Total Time Delay	3.0 + 0.5 seconds

Employment Distance:

Hand Thrown	50 feet (15.2 m)
Shotgun Launched	200 feet (61 m)

Projectile Specifications:

Number	100 (minimum)
Material	Rubber
Hardness	Durometer 70 to 80 on Shore "A" scale
Diameter	0.25 inches (6.4 mm) (minimum)

- Fuze separates from the body 1.3 to 1.7 seconds after primer ignition.
- Second delay causes burster charge detonation 3.0 + 0.5 seconds average after primer ignition. Rubber balls are then dispersed in a 50-feet (15-m) circular pattern.
- 12-gauge shotgun with launch cup and launch cartridge can launch grenade 200 feet (61 m) + 25 feet (7.6 m)

Figure B-6. Rubber Ball Grenade

Practice Grenade Body

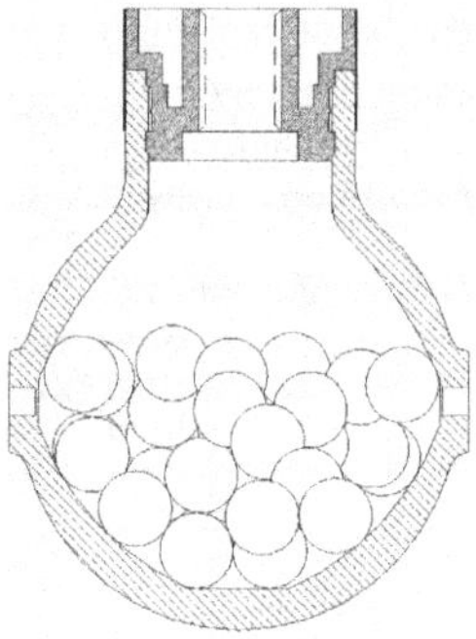

Using Services: USMC

Status: Fielded 2002

DODIC: GG05

NSN: Pending

Acquisition Activity: MARCORSYSCOM – NSWC Crane, IN

Further Information:

Commercial: MARCORSYSCOM (703) 784-9395
NSWC Crane (812) 854-5801

Diameter .... 3.12 inches (79 mm)
Projectiles .... N/A
Height .... 4.2 inches (106 mm)
Total Weight .... 208 grams
Explosive Weight .... N/A (inert)
Total Time Delay .... 3.0 + 0.5 seconds when used with practice grenade fuze

Employment Distance:

Hand Thrown .... 50 feet (15.2 m)
Shotgun Launched .... 200 feet (61 m) using Mossberg launch cup and 12-gauge launch cartridge (AA30).

- Practice grenade body is reusable up to 25 times and is a component of a practice grenade assembly, consisting of the body and a practice grenade fuze.
- Practice grenade assembly accurately replicates the GG04 rubber ball grenade and is used for training purposes.

Figure B-7. Practice Grenade Body

Practice Grenade Fuze

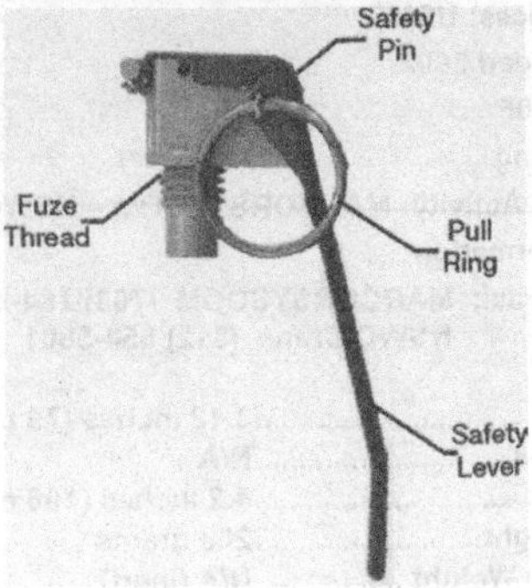

Using Services: USMC
Status: Fielded 2002
DODIC: Pending
NSN: Pending
Acquisition Activity: MARCORSYSCOM—NSWC Crane, IN
Further Information:
Commercial: MARCORSYSCOM (703) 784-9395
NSWC Crane (812) 854-5801

Color	Blue safety lever with brown stripe
Projectiles	N/A
Height	4.25 inches (107 mm)
Total Weight	82 grams
Explosive Weight	Not available
Total Time Delay	3.0 + 0.5 seconds

Employment Distance when used with GG05 Practice Grenade Body:

Hand Thrown	50 feet (15.2 m)
Shotgun Launched	200 feet (61 m)—using Mossberg launch cup and 12-gauge launch cartridge (AA30).

- Employ practice grenade fuze with the practice grenade body to create a practice grenade assembly consisting of the body and a practice grenade fuze.
- Practice grenade assembly accurately replicates the GG04 rubber ball grenade and is used for training purposes.

Figure B-8. Practice Grenade Fuze

Nonlethal Grenade Launch Cup

Using Service: USMC

Status: Fielded 2002

DODIC: N/A

NSN: Pending

Acquisition Activity: MARCORSYSCOM—NSWC Crane, IN

Further Information: Commercial: MARCORSYSCOM
(703) 784-2006 Ext. 2733
NSWC Crane (812) 854-5801

NOTE: Cup body shown is representative of cup bodies used on all shotguns. Adapter shown is subject to change depending on the shotgun used and design improvements.

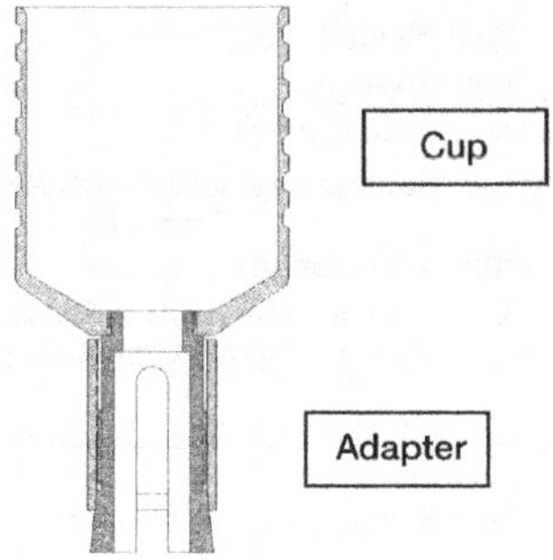

Cross Section View

Color .... Black
Diameter (internal) .... 3.12 inches (79 mm)
Length (without adapter) ..... 4 inches (102 mm)
Weight .... 397 grams (maximum)

- Use launch cup in concert with AA30 launch cartridge and either the GG04 rubber ball grenade or GG05 practice grenade assembly
- Attach launch cup and remove from the shotgun without using tools
- Launch cup can withstand accidental detonation of rubber ball grenade without fragmenting

Figure B-9. Nonlethal Grenade Launch Cup

Diversionary Charge, MK 141, MOD 0

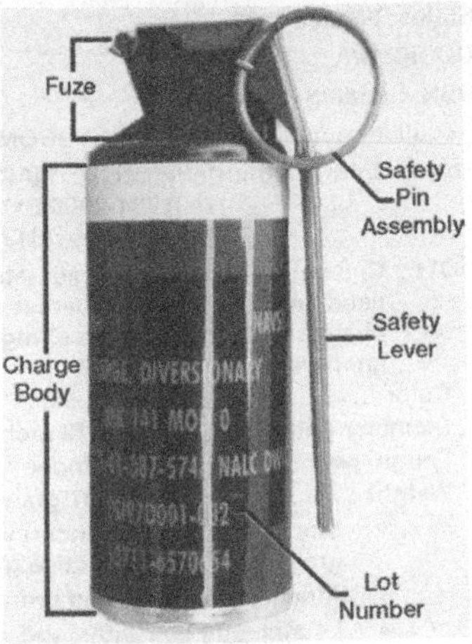

Using Services: USN, USMC
Status: Fielded 1997
DODIC: DWBS
NSN: 1375-01-387-5748
Acquisition Activity: MARCORSYSCOM—NSWC Crane, IN
Further Information:
Commercial: MARCORSYSCOM (703) 784-9395
NSWC Crane (812) 854-3934

Color	Brown with white stripe
Diameter	1.77 inches (45 mm)
Projectile	N/A
Length	5.214 inches (132.4 mm)
Total Weight	99.8 grams
Net Explosive Weight	17.8 grains
Light Output	2 million candlepower for 4 milliseconds
Noise Output	176 to 185 decibels
Material	Molded and machined polyurethane
Time Delay:	
Initial	1.26 seconds (minimum)
Total	1.34 to 1.84 seconds

- Diversionary charge can cause serious injury if held at time of detonation
- Light and noise most effective in confined space to stun individuals
- Most often employed for hostage rescue

Figure B-10. Diversionary Charge, MK 141, MOD 0

Cartridge, 12-Gauge Rubber Sting Ball™

Using Services: USCG

Status: Fielded

DODIC: N/A Model No. 2552

NSN: Commercial Procurement

Acquisitions Activity: Commandant (G-OCU)

Further Information: Commandant (G-OCU) (202) 267-1522

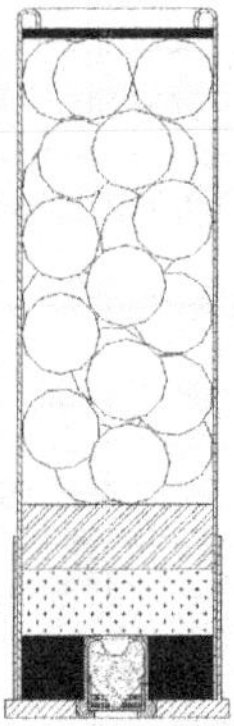

Model Number ....31-caliber sting-ball
Type ....Multiple projectile rubber pellets
Range ....10 yards (9 m)
Projectiles ....18 (average)
Cartridge Material.................Plastic
Projectile Weight ....35 grams each
Projectile Material..................31-caliber rubber pellets
Overall Height........................2.75 inches (70 mm)
Overall Weight ....18 grams
Compatibility..........................12-gauge shotguns
Velocity..................................400 feet/second

USCG

- Fatal injuries are possible at employment distances of less than 5 yards.
- At distances of 20 feet (6 m) to 40 feet (12 m), target area should be center mass
- This round is most effective against area targets with a maximum effective range of 15 yards
- Training needs to reinforce that head shots are NOT ACCEPTABLE
- Round is most effective against area targets (multiple personnel)
- Stock Sock: a yellow "stock sock" will be used whenever the modified 12-gauge Remington 870 shotgun is loaded with nonlethal munitions. This removable nylon sleeve is used to identify the weapon as a nonlethal delivery system.

- All specifications are average and are subject to change without notice
- 2552SPEC

Figure B-11. Cartridge, 12-Gauge Rubber Sting Ball™

Cartridge, 12-Gauge Sting Ball™ High Velocity

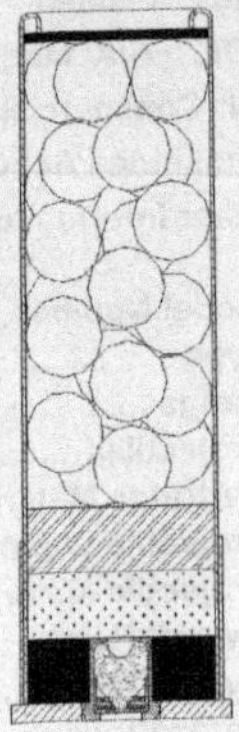

Using Services: USCG
Status: Fielded
DODIC: N/A Model No. 2553
NSN: Commercial Procurement
Acquisitions Activity: Commandant (G-OCU)
Further Information: Commandant (G-OCU) (202) 267-1522

Model Number	31-caliber sting-ball
Type	Multiple projectile rubber pellets
Range	10 yards (9 m)
Projectiles	18 (average)
Cartridge Material	Plastic
Projectile Weight	.35 grams each
Projectile Material	.31-caliber rubber pellets
Overall Height	2.4 to 2.5 inches (61 – 63 mm)
Overall Weight	18 grams
Compatibility	12-gauge shotguns
Velocity	900 feet/second

USCG

- Fatal injuries are possible at employment distances of less than 5 yards.
- At distances of 20 feet (6 m) to 40 feet (12 m), target area should be center mass
- Maximum effective range is 15 yards
- Training needs to reinforce that head shots are NOT ACCEPTABLE
- Round is most effective against area targets (multiple personnel)
- Stock Sock: a yellow "stock sock" will be used whenever the modified 12-gauge Remington 870 shotgun is loaded with nonlethal munitions. This removable nylon sleeve is used to identify the weapon as a nonlethal delivery system.
- All specifications are average and are subject to change without notice.
- 2553SPEC

Figure B-12. Cartridge, 12-Gauge Sting Ball™ High Velocity

M1006, 40-mm Nonlethal Cartridge (Sponge Grenade)

Using Services: U.S. Army

Status: Materiel Released April 2000

DODIC: BA06

NSN: 1310-01-452-1190

Acquisition Activity: Acquisition Center, Picatinny Arsenal, NJ

Further Information:

Commercial: Picatinny Arsenal NL Systems Integrator: (973) 724-6948
Product Director: (973) 724-6283

Color:

Projectile .... Black, olive green (foam rubber)
Cartridge Case .... Opaque
Diameter .... 40 millimeter
Length .... 5.272 inches (13.4 cm) (maximum)
Total Weight.......................... 68 grams (0.15 lbs)
Propellant Weight................. 2 grains
Muzzle Velocity..................... 265 + 20 feet/second (81 + 6 m/sec)
Acceptance Accuracy.......... >85 % accuracy at 30 meters (E-silhouette) and >60 % accuracy at 50 meters (E-silhouette) when fired from an M16A2/M203 or a M4/M203
Engagement Range............. 10 meters to 50 meters
Projectile Specification:
Shape .... 40-mm bullet shaped foam rubber
Material............................... Foam rubber
Weight .... 28.5 grams (.06 lbs)

- Fatal injuries are possible at employment distances of less than 32.8 feet (10 m)
- At distances of 10 meters to 50 meters, target area should be center mass
- Training needs to reinforce that head shots are NOT ACCEPTABLE
- Do not skip fire this round! Round becomes unpredictable upon striking the ground
- Round is most effective against individually selected targets (point round)

Figure B-13. M1006, 40-mm Nonlethal Cartridge (Sponge Grenade)

M1012, 12-Gauge Shotgun: Nonlethal, Point Control Cartridge

Using Services: U.S. Army
Status: Type Classified, June 2000
DODAC: AA51
NSN: 1305-01-470-2405
Acquisition Activity: Acquisition Center, Picatinny Arsenal, NJ
Further Information:
Commercial: Picatinny Arsenal NL Systems Integrator: (973) 724-6948
Product Director: (973) 724-6086

Color:
Cartridge Case: ....Clear
Diameter................................12 gauge
Length ....2.45 inches (62.2 mm)
Total Weight...........................500 grains
Propellant Weight.................6.6 grains smokeless (Red Dot)
Muzzle Velocity.....................500 feet/second (152.4 m/sec)
Acceptance Accuracy..........80 % accuracy at 10 and 20 meters (E-silhouette) fired out of Mossberg 500, Mossberg 590, and Winchester 1200
Engagement range.............10 to 20 meters
Projectile Specification:
Shape ....Fin stabilized 'bomblet' shape
Material.............................Rubber
Weight ....0.2 ounces (5.8 gm)
Hardness.......................... 75 Durometer "A" Scale

- Fatal injuries are possible at employment distances of less than 33 feet (10 m)
- At distances of 10 to 20 meters, target area should be center mass
- Training needs to reinforce that head shots are NOT ACCEPTABLE
- Round is most effective against individually selected targets (point round)

Figure B-14. M1012, 12-Gauge Shotgun: Nonlethal, Point Control Cartridge

M1013, 12-Gauge Shotgun: Nonlethal, Crowd Dispersal Cartridge

Using Services: U.S. Army

Status: Type Classified June 2000

DODAC: AA52

NSN: 1305-01-470-2139

Acquisition Activity: Acquisition Center, Picatinny Arsenal, NJ

Further Information:

Commercial: Picatinny Arsenal NL Systems Integrator: (973) 724-6948
Product Director :(973) 724-6086

Color:
- Projectile .... Black rubber
- Cartridge Case .... Clear

Diameter .... 12 gauge
Projectile .... 18 rubber balls
Length .... 2.37 inches (60.2 mm)
Total Weight........................... 500 grains (32.4 gm)
Propellant Weight................. 14 grains smokeless (Red Dot)
Muzzle Velocity...................... 900 feet/second (274 m/sec)
Acceptance Accuracy.......... 80 % accuracy for a 2-by-3-meter target at 10 and 20 meters fired out of Mossberg 500, Mossberg 590, and Winchester 1200
Engagement range............... 10 to 20 meters
Projectile Specification:
- Shape................................ 32-gauge (0.32 in) ball
- Material............................. PVC rubber compound
- Weight................................ 0.01 ounces (0.25 gm)
- Hardness .... 75 Durometer "A" Scale

- Fatal injuries are possible at employment distances of less than 32.8 feet (10 m)
- At distances of 10 to 20 meters, target area should be center mass
- Training needs to reinforce that head shots are NOT ACCEPTABLE

Figure B-15. M1013, 12-Gauge Shotgun: Nonlethal, Crowd Dispersal Cartridge

M1029, 40-mm Nonlethal Crowd Dispersal Grenade

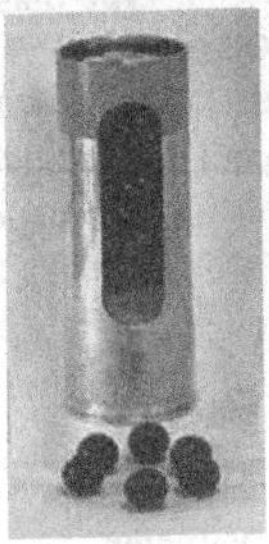

Using Services: U.S. Army
Status: Type Classified May 2001
DODIC: BA13
NSN: 1310-01-475-0628
Acquisition Activity: Acquisition Center, Picatinny Arsenal, NJ
Further Information:
Commercial: Picatinny Arsenal NL Systems Integrator: (973) 724-6948
Product Director: (973) 724-6283

Color:
- Projectile .... Black rubber
- Cartridge Case .... Aluminum

Diameter .... 40 millimeter
Projectile .... 48 rubber balls
Length .... 4.8 inches (12.2 cm)
Total Weight .... 202.22 grams (7.0 oz)
Propellant Weight .... 8.5 grains smokeless
Muzzle Velocity .... 450 feet/second
Acceptance Accuracy .... > 90 % hit accuracy 4 out of 5 (shoulder to shoulder) E-silhouettes target at 30 meters and when fired from an M16A2/M203 or a M4/M203

Projectile Specification:
- Shape .... 48-gauge ball (.48 in diameter)
- Material .... Rubber
- Weight .... 1.3 grams (.046 oz)
- Hardness .... Durometer 60, scale "A"

- Fatal injuries are possible at employment distances of less than 32.8 feet (10 m)
- At distances of 10 to 30 meters, target area should be center mass of group of individuals
- Round is most effective when used to disperse groups of individuals

Figure B-16. M1029, 40-mm Nonlethal Crowd Dispersal Grenade

L96A1*, Grenade, Discharger: Antiriot, Irritant, CS

Using Services: U.S. Army, USMC

Status: Materiel Release, 4 January 2002

DODIC: FZ14

NSN: 1330-01-459-4018

Acquisition Activity: Acquisition Center, Picatinny Arsenal, NJ

Further Information:

Commercial: Picatinny Arsenal NL Systems Integrator: (973) 724-6948
Project Officer: (410) 436-2139

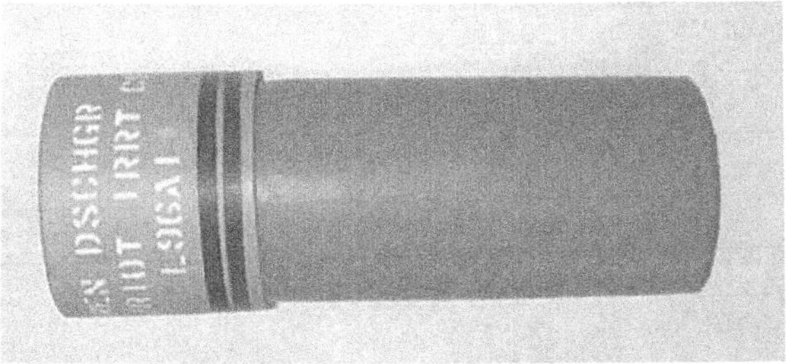

Color ....Light gray with red and brown bands
Diameter ....66 millimeters
Projectile ....23 canisters filled with CS compound
Length ....185 millimeters (7.28 in)
Total Weight..........................568 grams (1.25 lbs)
Propellant Weight................. 3.8 grams (.008 lbs)
Muzzle Velocity...................... 35.8 meters/second (117.5 ft/s)
Acceptance Accuracy..........65 to 95 meters
Deployment Range...............65 to 95 meters
Projectile Specification:

NOTE: It is not accurate to refer to the submunitions of the L96A1 grenade as projectiles. The antiriot material (CS) is released from the submunitions as they burn on the ground.

Shape...............................Cylindrical
Material ....Aluminum
Weight..............................11.3 grams (.025 lbs)

* The L97A1 Grenade is the training grenade for the L96A1 Grenade.

Figure B-17. L96A1, Grenade, Discharger: Antiriot, Irritant, CS

L97A1, Grenade, Discharger: Antiriot, Practice

Using Services: U.S. Army, USMC

Status: Material Released, 4 January 02

DODIC: FZ15

NSN: 1330-01-459-4032

Acquisition Activity: Acquisition Center, Picatinny Arsenal, NJ

Further Information:

Commercial: Picatinny Arsenal NL Systems Integrator: (973) 724-6948 Project Officer: (410) 436-2139)

Color......................................Blue with brown and green bands
Diameter................................66 millimeters
Projectile...............................23 canisters filled with Cinnamic Acid (CA)
Length...................................185 millimeters (7.28 in)
Total Weight..........................568 grams (1.25 lbs)
Propellant Weight.................3.8 grams (.008 lbs)
Muzzle Velocity.....................35.8 meters/second (117.5 ft/s)
Acceptance Accuracy..........65 to 95 meters
Deployment Range...............65 to 95 meters
Projectile Specification:

NOTE: It is not accurate to refer to the submunitions of the L97A1 grenade as projectiles. The antiriot, practice material (CA) is released from the submunitions as they burn on the ground.

Shape...............................Cylindrical
Material............................Aluminum
Weight..............................11.3 gram (.025 lbs)

- CA smoke can cause irritation to the eyes, mucous membranes, and skin. Personnel entering a CA cloud must wear appropriate face protection such as an M17- or M40-series protective mask.
- The burning grenade canisters could start a fire in dry grasses. Keep fire extinguisher available.
- During firing training, personnel within 16 meters must wear single hearing protection. This includes personnel within vehicle.
- Do not fire grenades when personnel or equipment are within 150 meters covering a 90-degree arc around a firing discharger.
- Do not fire grenades at the 50- or 75-meter position on the M315 adjustable aiming bracket. Fire only at the 100-meter setting.

Figure B-18. L97A1, Grenade, Discharger: Antiriot, Practice

Grenade, Launcher: Nonlethal, Distraction, M98

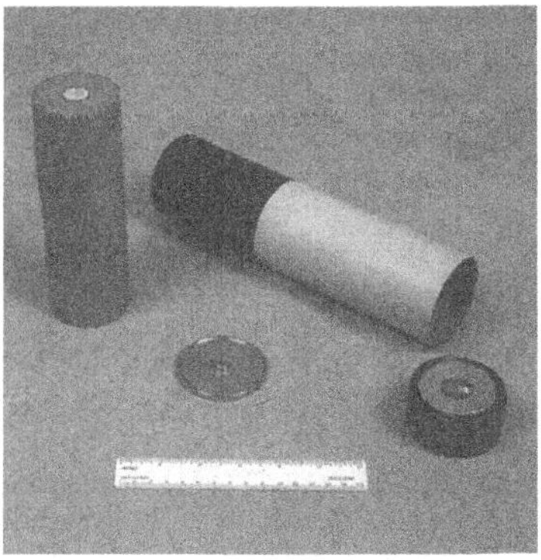

Using Services: U.S. Army, USMC
Status: Type Classified, 17 October 2001
DODIC: FZ16
NSN: 1330-01-484-7773
Acquisition Activity: Acquisition Center, Picatinny Arsenal, NJ
Further Information: Commercial: Picatinny Arsenal
NL Systems Integrator:
(973) 724-6948
Project Officer: (410) 436-2139

Color ....Green and black with brown band
Diameter ....66 millimeters
Projectile ....3 bursting submunitions, with pyrotechnic charges for audio and visual stimuli
Length ....25.25 centimeters (9.94 in)
Total Weight...........................725.7 grams (1.6 lbs)
Propellant Weight.................5.0 grams (.011 lbs)
Muzzle Velocity......................48.9 meters/seconds (160.4 ft/s)
Acceptance Accuracy..........100 to 120 meters
Deployment Radius..............10 to 15 meters
Projectile Specification:

NOTE: It is not accurate to refer to the submunitons of the M98 grenade as projectiles. These submunitions are ground bursting and the payload of inert ballast is just for flight stabilization and to deliver the audio/visual stimuli to the required range.

Filler ....Inert Ballast

- There is a remote chance of starting a fire. Keep fire extinguisher available.
- Personnel within 17 meters of a firing discharger must wear single hearing protection. This includes personnel within vehicle.
- Avoid looking directly at a bursting grenade; it could result in temporary loss of vision, lasting 2 to 5 minutes.
- Personnel within 35 meters of an operating launcher or 0.5 meters of a bursting canister should wear safety or ballistic-type eye protection, military clothing with long sleeves and elevated shirt collar, and helmet headgear.
- Do not handle or load M98s within 50 meters of a frequency transmitter emitting at 175 MHz, 448 MHz, 503 MHz and 1430 MHz.
- Do Not fire grenades when personnel or equipment are within 160 meters covering a 90-degree arc around a firing discharger.

Figure B-19. Grenade, Launcher: Nonlethal, Distraction, M98

M99, Grenade, Launcher: Nonlethal, Blunt Trauma

Using Services: U.S. Army, USMC
Status: Type Classified—17 October 2001
DODIC: FZ17
NSN: 1330-01-484-7775
Acquisition Activity: Acquisition Center, Picatinny Arsenal, NJ
Further Information:
Commercial: Picatinny Arsenal NL Systems Integrator: (973) 724-6948
Project Officer: (410-436-2139)

Color....................................Green and black with brown band
Diameter................................66 millimeters
Projectile...............................3 bursting submunitions, each with 140 balls
Length...................................25.25 centimeters (9.94 in)
Total Weight..........................725.7 grams (1.6 lbs)
Propellant Weight..................5.0 grams (.011 lbs)
Muzzle Velocity......................48.9 meters/second (160.4 ft/s)
Acceptance Accuracy..........100 to 120 meters
Deployment Radius..............5 to 10 meters
Projectile Specification:

NOTE: It is not accurate to refer to the submunitions of the M99 grenade as projectiles. These submunitions are ground bursting and the payload of inert ballast is just for flight stabilization and to deliver the audio/visual stimuli to the required range.

Shape..............................32-caliber (5/16 in) ball
Material...........................PVC
Weight.............................0.40 grams (.014 oz)
Hardness.........................90 Durometer

- There is a remote chance of starting a fire. Keep fire extinguisher available.
- Personnel within 17 meters of a firing discharger must wear single hearing protection. This includes personnel within vehicle.
- Avoid looking directly at a bursting grenade; it could result in temporary loss of vision, lasting 2 to 5 minutes
- Personnel within 35 meters of an operating launcher or 0.5 meters of a bursting canister should wear safety or ballistic-type eye protection, military clothing with long sleeves and elevated shirt collar, and helmet headgear.
- Do not handle or load M98s within 50 meters of a frequency transmitter emitting at 175 MHz, 448 MHz, 503 MHz and 1430 MHz.
- DO NOT fire grenades when personnel or equipment are within 160 meters covering a 90-degree arc around a firing discharger.
- To minimize risk of eye injury, use the M99 primarily against crowds of adults.

Figure B-20. M99, Grenade, Launcher: Nonlethal, Blunt Trauma

Appendix C

Interservice Nonlethal Individual Weapons Instructor Course

1. Introduction

This appendix provides an overview of the Interservice Nonlethal Individual Weapons Instructor Course (INIWIC). The INIWIC is a "train-the-trainer" course taught in 15 subcourses and divided into four sections. Section I provides INIWIC descriptive data that includes the purpose, scope, peacetime and mobilization training requirements, student prerequisites, and equipment/ammunition requirements necessary to successfully complete the course. Section II lists the total training days and hours required per academic subcourse. Section III gives a detailed description of each subcourse and terminal learning objectives that each student must meet to complete the course. Section IV is a listing of 15 annexes that comprise the INIWIC Program of Instruction (POI). Access the entire INIWIC, with history, updates, and lesson plans at the following Internet address: http://www.wood.army.mil/mcdet/niwic/niwic.htm.

2. Course Descriptive Data

a. Course Title: Interservice Nonlethal Individual Weapons Instructor Course

b. Location: Marine Corps Detachment, Fort Leonard Wood, Missouri

c. Marine Corps service school code: A16H5A3

d. Other service course number: USAF—L5AZA3P071019

e. Military assistance program articles and service list number: P166810

f. Purpose: To certify selected DOD/DOT service persons as nonlethal individual weapons instructors.

g. Scope: This is a train-the-trainer course designed for E-4s and above. The objective of this course is to prepare students to train individuals and groups in the utilization and employment of NLW and to serve as the unit NLW subject matter expert. This course is designed for all occupational specialties

h. Length: 10 training days

i. Curriculum Breakdown: 100 total hours.

(1) 2.00 Demonstration.

(2) 24.00 Demonstration/practical exercise.

(3) 22.50 Lecture.

(4) 5.00 Lecture/demonstration.

(5) 4.00 Live-fire exercise.

(6) 26.50 Practical application.

(7) 6.00 Performance exam.

(8) 4.00 Written exam.

(9) 6.00 Administrative time.

j. Maximum Class Capacity: 30.

k. Optimum Class Capacity: 25.

l. Minimum Class Capacity: 15.

m. Class Frequency: 12.

n. Prerequisites: Students must be selected by commanders to be nonlethal weapons instructors and be an E-4 or above.

o. MOS Received: None.

p. Quota Control: Formal Schools Branch, Training and Education Command, Quantico, Virginia.

q. Funding: Service or command funded.

r. Reporting Instructions: All students report with orders to the Fort Leonard Wood, Missouri, Billeting Office, Bldg # 470 no later than 2400 on the day prior to class convening. Government quarters are available; however, messing is not.

s. Instructor Staffing: Two USMC instructors and two U.S. Army instructors.

t. School Overhead: Course OIC and chief instructor

Training Support: Training support is provided by the USMC for all common training support items. Service-specific training items are provided by the respective service.

Table C-1. Required Equipment For INIWIC

Item	***Quantity***
Personnel Protectors:	
Non-Ballistic Riot Face Shield	40
Non-Ballistic Full Length Riot Shield	40
Non-Ballistic Shin Guards	40
Ballistic Riot Face Shield	20
Ballistic Full Length Riot Shield	20
Ballistic Shin Guards	20
Kevlar Helmets	40
Flak Jacket/Body Armor	40
Personnel Effectors:	
Disposable Handcuffs (Flex Cuffs)	300
12-Gauge Area (rubber ball)	1400
12-Gauge Beanbag	1400
12-Gauge Fin Stabilized	1400
12-Gauge Flash-bang	1400
12-Gauge OO Buck	xxxx
12-Gauge 7 1/2 shot	xxxx
12-Gauge Launching Cup Cartridge	1400
40-mm Area (rubber ball)	1400
40-mm Foam Baton (3 foam sub-munitions)	1400
40-mm Sponge Grenade	1400
MK141 Diversionary Grenade	600

Table C-1. Required Equipment For INIWIC

Item	*Quantity*
M84 Reloadable Diversionary Grenade	600
Sting Ball Hand Thrown or Shot launched Grenade	600
Mossberg 500A2 Shotgun Launch Cup	15
M203 Grenade Launcher	15
Mossberg 500 3" Shotgun	15
36" Riot Oak Baton w/Belt Ring	40
Monadnock® 24-36" Expandable Baton w/holster	40
M36 RCA Dispenser	40
Mk-4 Personal OC Dispenser	40
Mk-9 Team OC Dispenser	40
Mk-46 High Capacity High Output OC Dispenser	20
MK-46 Hardware Kit (Refill Kit)	10
M33A1 Squad Riot Control Agent Dispenser	10
Modular Crowd Control Munitions Ground Emplaced (MCCM-GE)	20
66-mm Vehicle Launched Nonlethal Grenade (VLNLG)	3
66-mm NL munitions	30
Mission Enhancers:	
Portable Bullhorn	5
40-mm Carrying Pouch	40
Diversionary Rubber Ball Grenade Pouch	40
12-Gauge Butt Cuff carriers	40
12-Gauge Utility 25-Round Pouch	40
Mk-4 Individual Carrying Pouch	40
Portable Public Address System w/Microphone	3
12-Volt Hand-Held Spotlight	5
Night Hunter High Intensity Light System w/Kit	5
12-Gauge High Intensity Light Set	5
Caltrops	5
Road Side Spike Strip	5
Portable Vehicle Arrest Barrier (PVAB)	3
Training Devices:	
Impact Training Suit	6
Striking Bags	15
Training Batons (24 in)	40
Training Batons (36 in)	40
Practice Grenade Body	40
Practice Grenade Fuse	540
• OPFOR Training Munitions (foreign objects)	xxxx
• Target Requirements (full or partial torso)	40
MK-4 Personal OC Dispenser (training)	300
MK-9 Team OC Dispenser (training)	300
Training Disposable Handcuffs (flex cuffs)	100
12-Gauge Dummy Rounds	100

Table C-2. INIWIC Class V Requirements

Type	*#Per Student*	*# Demo Rounds*	*Total Per Class*	*Total Per Year*
12-Gauge Area (rubber ball)	5	5	155	1400
12-Gauge Beanbag	5	5	155	1400
12-Gauge Fin Stabilized	5	5	155	1400
40-mm Area (rubber ball)	5	5	155	1400
40-mm Foam Baton	5	5	155	1400
Sting Ball Hand Thrown or Shotgun Launched Grenades	2	2	62	600
MK141 Diversion Devices	2	2	62	600

3. Training Days Summary

This course consists of 10 training days. The curriculum is broken down into three categories.

Table C-3. INIWIC Curriculum

NLW Tactics	***Hours***
Force Continuum/ROE	2.00
Crowd Dynamics/Control/Formations	14.00
Communication Skills	2.00
MOOTW	2.00
Tactical Considerations	2.00
	22.00
NLW Systems	***Hours***
Oleoresin Capsicum	4.00
Empty Hand Control Techniques	19.00
Riot Baton Techniques	8.00
Nonlethal Munitions and Equipment	8.00
Portable Vehicle Barrier	5.00
66-mm Grenade Launcher	2.00
Modular Crowd Control Munition	2.00
	48.00
Nonlethal Weapons Instructor	***Hours***
Techniques of Military Instruction	10.00
NLW Scenario Exercise	10.00
	20.00
Subtotal =	**90.00**
Examinations =	***4.00***
Administration =	***6.00***
Total Hours =	***100.00***

4. Course Body

The course body consists of 15 subcourses. A quick synopsis of each subcourse is included below.

a. **Instructor Development**. This subcourse teaches and reinforces basic military instructor methodology. It identifies characteristics of good instructors and four ways they can improve their teaching ability. Additionally, it illustrates rules of conducting military instruction as well as the steps in planning a lesson and teaching classes.

b. **Force Continuum**. This subcourse introduces the five levels of resistance as well as identifying the levels of the force continuum. It illustrates influencing the use of force decisions and the definitions relating to deadly force. Lastly, it identifies the three preconditions for deadly force and seven authorized occasions for use of deadly force.

c. **Crowd Dynamics/Crowd Control**. This subcourse outlines the differences between crowds, mobs, and riots, and teaches the student basic crowd control techniques to apply in various situations. Upon completion, the student will be able to instruct others in both the behavioral aspects of a crowd during a civil disturbance and also various techniques of controlling a civil disturbance. It will identify factors that affect both individual and crowd behavior and it identifies all crowd control equipment needed, or essential, for mission accomplishment.

d. **Empty Hand Control Techniques**. This subcourse identifies safety factors to consider when instructing empty hand control techniques by illustrating the four Ms (mind, mass, movement, muscle) of self-defense, psychomotor skills, and phases of psychomotor skills. It illustrates levels of training and identifies common pressure points of the human body. It identifies hinges, appropriate angles of approach, and recommended patterns of movement. Instructors will learn blocking techniques, striking skills, control holds, escort positions, and restraint devices.

e. **Baton Techniques**. This subcourse identifies riot baton nomenclature and safety standards when conducting riot baton training. It identifies methods of teaching psychomotor skills and the phases of teaching psychomotor skills. It highlights baton-striking techniques, blocking techniques, and retention techniques.

f. **Military Operations Other Than War (MOOTW)**. This subcourse defines operations other than war and the difference between MOOTW and declarations of war. It helps identify the primacy of political objectives and the strategic aspects of MOOTW. It illustrates the range of operations and the principles of operations, as well as the types of military operations other than war.

g. **Rules of Engagement (ROE**). This subcourse defines ROE and specific terms that apply to these rules. It also defines the purpose of ROE and provides the student an understanding of policies underlying the need for ROE. It identifies three sources for ROE and the two parts of ROE that will not change. It explains how ROE relate to using and applying NLW.

h. **Oleoresin Capsicum (OC) Training**. This subcourse explains how the three levels of OC are created, as well as the nomenclatures for all aerosol canisters. It highlights the different spray patterns, methods of delivery, and cautions concerning the traumatic hydraulic needle effect. It explains employment considerations and the three levels of OC contamination. Additionally, it explains the physical and mental effects of OC and trains students on different grips, draw methods, and proper stances when employing OC. It explains procedures for decontamination, recovery, and first aid.

i. **Types of Formations**. This subcourse explains the elements and individual responsibilities within a riot control formation. Additionally, it trains proper grip, balance, holding, and retention of the riot control shield. It highlights leadership requirements by teaching different commands and cadence for conducting riot control formations. It illustrates the different types of formations and the support elements of a riot control operation

j. **Tactical Considerations**. This subcourse addresses the preparations necessary for crowd control operations and the graduated response of a crowd control force. It identifies support and information required from host nation police and civil authorities. It describes the employment measures of military forces in crowd control and MOOTW environments. It highlights the actions, optional measures, and guidance for using NLW in crowd control and MOOTW environments.

k. **Nonlethal Munitions and Employment**. This subcourse identifies the types of nonlethal munitions and their intended purpose. It illustrates how nonlethal munitions fall within established "use of force" continuums. It identifies the types, characteristics, and ranges of all currently employed nonlethal munitions.

l. **Communications Skills**. This subcourse identifies the areas of nonverbal communication employed to determine an individual's state of mind. This, along with an understanding of the importance of personal space, gives the student the tools to resolve conflict with the amount of force as applied in the force continuum.

m. **Portable Vehicle Arrest Barrier**. This subcourse teaches the employment and maintenance of a force protection asset assigned to the nonlethal capability set. The PVAB is designed to stop a 7,500-pound vehicle going 45 mph. The PVAB protects gates and check points.

n. **Modular Crowd Control Munition**. This subcourse teaches the employment of the MCCM; the nonlethal version of the claymore mine. The MCCM is a force protection asset designed to help protect static defensive positions or check points.

o. **M-315 66-mm Grenade Launcher**. This subcourse teaches the installation, employment, and maintenance of the M-315. The M-315 increases the current maximum range of nonlethal sting ball munitions from approximately 50 meters to 100 meters. Additionally, this weapons platform can deploy flash bangs, smoke, and CS munitions.

Appendix D

Sample: Unit Training Schedules/Firing Tables

Nonlethal Individual Weapons Training Course Training Schedule: Unit Training Schedule Number 1.

Training Day #1					
Time	***Subject***	***Location***	***Uniform***	***Instr***	***Ref/Notes***
0800	Administration Time	Classroom	U1		
0830	Force Continuum	Classroom	U1		
1000	ROE	Classroom	U1		
1200	Chow	Mess Hall	U1		
1300	MOOTW	Classroom	U1		
1500	Communication Skills	Classroom	U1		
1700	Secure/COB				

Administration:

Uniforms:

U1: BDU/utility uniform
U2: BDU/utility uniform with LBE/harness and canteens
U3: BDU/utility uniform with LBE/harness, canteens, flak jacket, and helmet
U4: BDU/utility uniform with LBE/harness, canteens, flak jacket, helmet, and assigned weapon

Transportation:
Weapons:
Ammunition:
Required Equipment:
Screen
Video projector with central processing unit
Televisions
VCR
Miscellaneous:
All personnel are required to have a notebook, pencil, and pen.
Handouts prepared and passed out.

Training Day #2					
Time	***Subject***	***Location***	***Uniform***	***Instr***	***Ref/Notes***
0800	Nonlethal Equipment Brief	Classroom	U1		
0900	Pressure Point Control Techniques/ Open Hand Control Techniques	Classroom	U1		
1000	Pressure Point Control Techniques/ Open Hand Control Techniques (Practical Application)	PT Field	U2		
1200	Chow				
1300	Pressure Point Control Techniques/ Open Hand Control Techniques (Practical Application)	PT Field	U2		
1700	Debrief/Secure/COB				

Administration:

Uniforms:
U1: BDU/utility uniform
U2: BDU/utility uniform with LBE/harness and canteens
U3: BDU/utility uniform with LBE/harness, canteens, flak jacket, and helmet
U4: BDU/utility uniform with LBE/harness, canteens, flak jacket, helmet, and assigned weapon

Transportation:

Weapons:

Ammunition:

Required Equipment:
Screen
Video projector with central processing unit
Flexicuffs (one set of cuffs per two students)
Riot Training Bags (one bag per two students)

Miscellaneous:
All personnel are required to have a notebook, pencil, and pen.
Handouts prepared and passed out.
Medical personnel on station.

Training Day #3					
Time	***Subject***	***Location***	***Uniform***	***Instr***	***Ref/Notes***
0800	Pressure Point Control	Classroom	U1		
	Techniques/Open Hand Control Techniques	Classroom	U1		
0900	Pressure Point Control Techniques/Open Hand Control Techniques (Practical Application)	PT Field	U1		
1200	Chow				
1300	Pressure Point Control Techniques/Open Hand Control Techniques (Practical Application)	PT Field	U2		
1700	Debrief/Secure/COB				

Administration:

Uniforms:
U1: BDU/utility uniform
U2: BDU/utility uniform with LBE/harness and canteens
U3: BDU/utility uniform with LBE/harness, canteens, flak jacket, and helmet
U4: BDU/utility uniform with LBE/harness, canteens, flak jacket, helmet, and assigned weapon

Transportation:

Weapons:

Ammunition:

Required Equipment:
Screen
Video projector with central processing unit
Flex cuffs (one set of cuffs per 2 students)
Riot Training Bags (one bag per 2 students)

Miscellaneous:
All personnel are required to have a notebook, pencil, and pen.
Handouts prepared and passed out.
Medical personnel on station.

Training Day #4					
Time	*Subject*	*Location*	*Uniform*	*Instr*	*Ref/Notes*
0800	Impact Weapons	Classroom	U1		
0900	Impact Weapons (Practical Application)	PT Field	U2		
1200	Chow				
1300	Impact Weapons (Practical Application)	PT Field	U2		
1700	Debrief/Secure/COB				

Administration:

Uniforms:
U1: BDU/utility uniform
U2: BDU/utility uniform with LBE/harness and canteens
U3: BDU/utility uniform with LBE/harness, canteens, flak jacket, and helmet
U4: BDU/utility uniform with LBE/harness, canteens, flak jacket, helmet, and assigned weapon

Transportation:

Weapons:
Baton (one baton per 2 students)
Training baton (one baton per 2 students)

Ammunition:

Required Equipment:
Screen
Video projector with central processing unit
Flex cuffs (one set of cuffs per 2 students)
Riot training bags (one bag per 2 students)
Three (3) riot training suits

Miscellaneous:
All personnel are required to have a notebook, pencil, and pen.
Handouts prepared and passed out.
Medical personnel on station.

Training Day #5					
Time	***Subject***	***Location***	***Uniform***	***Instr***	***Ref/Notes***
0800	Impact Weapons (Practical Application)	PT Field	U2		
1000	Crowd Dynamics/Crowd Control	Classroom	U1		
1130	Chow				
1230	Crowd Dynamics/Crowd Control	Classroom	U1		
1330	Riot Control Formations	Classroom	U1		
1430	Oleoresin Capsicum (OC) Aerosol Training	Classroom	U1		
1630	Debrief/Secure/COB				

Administration:

Uniforms:

U1: BDU/utility uniform

U2: BDU/utility uniform with LBE/harness and canteens

U3: BDU/utility uniform with LBE/harness, canteens, flak jacket, and helmet

U4: BDU/utility uniform with LBE/harness, canteens, flak jacket, helmet, and assigned weapon

Transportation:

Weapons:

Training baton (one baton per 2 students)

Baton (one baton per 2 students)

Ammunition:

Inert OC for classroom instruction

Required Equipment:

Riot training bags (one bag per 2 students)

Screen

Video projector with central processing unit

Three (3) riot training suits

Miscellaneous:

All personnel are required to have a notebook, pencil, and pen.

Handouts prepared and passed out.

Training Day #6					
Time	***Subject***	***Location***	***Uniform***	***Instr***	***Ref/Notes***
0800	OC Aerosol Training (Practical Application)	PT Field	U2		
1200	Chow				
1300	OC Aerosol Training (Practical Application)	PT Field	U2		
TBD	Debrief/Secure/COB				

Administration:

Uniforms:
U1: BDU/utility uniform
U2: BDU/utility uniform with LBE/harness and canteens
U3: BDU/utility uniform with LBE/harness, canteens, flak jacket, and helmet
U4: BDU/utility uniform with LBE/harness, canteens, flak jacket, helmet, and assigned weapon

Transportation:

Weapons:

Ammunition:
MK 4.inert OC (one per student)
MK 4 individual can OC (2 cans)

Required Equipment:
Five (5) large trash cans with liners
Two (2) water hoses with running water

Miscellaneous:
Medical personnel on station.

Training Day #7					
Time	***Subject***	***Location***	***Uniform***	***Instr***	***Ref/Notes***
0730	Movement to Range		U4		
0800	Nonlethal Munitions and Employment	Range Classroom	U1		
1030	Chow	Range			
1100	Safety Brief/Range Brief	Range	U4		
1130	Nonlethal Munitions (Live Fire)	Range	U4		
TBD	Evening Chow				
TBD	Safety Brief/Range Brief	Range	U4		
1800	Night Exercise Sun Set +30)	Range	U4		
TBD	Debrief/Clean Up	Range	U1		
TBD	Movement to Rear				
TBD	Secure/ COB				

Administration:

Uniforms:
U1: BDU/utility uniform
U2: BDU/utility uniform with LBE/harness and canteens
U3: BDU/utility uniform with LBE/harness, canteens, flak jacket, and helmet
U4: BDU/utility uniform with LBE/harness, canteens, flak jacket, helmet, and assigned weapon

Transportation: Yes

Weapons:
Assigned Weapons

Ammunition:
12-gauge rubber ball
12-gauge fin stabilized
12-gauge bean bag
12-gauge launching cup cartridge
40-mm foam baton
40-mm rubber ball
40-mm sponge grenade
Stingball grenades

Required Equipment:
Eight (8) 12-Gauge stinger launcher cup
Fifteen (15) targets
Eye and hearing protection
Fifteen (15) full-length riot shields
All students require munitions pouches, shin guards, face shields
Two (2) MRE per student
Video projector with central processing unit
Three (3) high-intensity light
Thirty (30) individual tactical lights

Miscellaneous:
Range safety officer.
Medical personnel on station.
Secondary means of communication (radios).
Conduct operational risk assessment.

Training Day #8					
Time	*Subject*	*Location*	*Uniform*	*Instr*	*Ref/Notes*
0700	Movement to Training Area #1		U2		
0730	Portable Vehicle Arresting Barrier (PVAB)	TA#1	U1		
0900	PVAB (Practical Application)	TA#1	U2		
1200	Chow				
1300	PVAB (Practical Application)	TA#1	U2		
1500	Debrief/Movement to Rear				
1530	Preventive Maintenance of PVAB				
1600	Secure/COB				

Administration:

Uniforms:
U1: BDU/utility uniform
U2: BDU/utility uniform with LBE/harness and canteens
U3: BDU/utility uniform with LBE/harness, canteens, flak jacket, and helmet
U4: BDU/utility uniform with LBE/harness, canteens, flak jacket, helmet, and assigned weapon

Transportation: Yes

Weapons:

Ammunition:

Required Equipment:
One (1) portable vehicle-arresting barrier (PVAB)
Video projector with central processing unit

Miscellaneous:
(1) MRE per student.

Training Day #9					
Time	*Subject*	*Location*	*Uniform*	*Instr*	*Ref/Notes*
0730	Movement to Range		U4		
0800	Modular Crowd Control Munition-Ground Emplaced (MCCM-GE)	Range Classroom	U1		
0900	Safety Brief/Range Brief	Range	U4		
0930	MCCM-GE (Practical Application	Range	U4		
1030	Chow				
1130	Vehicle Mounted Nonlethal Grenade Launcher (VMNLGL)	Range Classroom	U4		
1300	VMNGL	Range	U4		
1730	Debrief/Movement to Rear				
1800	Secure/COB				

Administration:

Uniforms:

U1: BDU/utility uniform
U2: BDU/utility uniform with LBE/harness and canteens
U3: BDU/utility uniform with LBE/harness, canteens, flak jacket, and helmet
U4: BDU/utility uniform with LBE/harness, canteens, flak jacket, helmet, and assigned weapon

Transportation: Yes

Weapons:

Vehicle Mounted Nonlethal Grenade Launcher (VMNLGL)

Ammunition:

(2) MCCM-GE, (2) 66-mm Riot Control Training Grenade (L97E1)

Required Equipment:

Eye and ear protection
Video projector with central processing unit

Miscellaneous:

Range safety officer.
Medical personnel on station.
Secondary means of communication.
Two (2) MRE per student.

Training Day #10					
Time	*Subject*	*Location*	*Uniform*	*Instr*	*Ref/Notes*
0700	Movement Down Range				U4
0730	Tactical Considerations	MOUT Classroom		U1	
0930	Safety Brief/Range Brief	MOUT Facility		U4	
1000	Tactical Exercise* (Practical Application)	MOUT Facility		U4	
1200	Chow				
1300	Tactical Exercise (Practical Application)	MOUT Facility		U4	
1700	Debrief/Movement to Rear				
1730	Weapons Maintenance				
1800	Secure/COB				

Administration:

Uniforms:
U1: BDU/utility uniform
U2: BDU/utility uniform with LBE/harness and canteens
U3: BDU/utility uniform with LBE/harness, canteens, flak jacket, and helmet
U4: BDU/utility uniform with LBE/harness, canteens, flak jacket, helmet, and assigned weapon

Transportation: Yes

Weapons:
Assigned weapon
Inert OC individual (one per student)
Training batons (one per student)
Fifty (50) practice grenades

Ammunition:

Required Equipment:
Three (3) riot training suit
All students require munitions pouches, flex cuffs, shin guards, face shields
Video projector with central processing unit
Fifteen (15) full length riot shields

Miscellaneous:
Range safety officer.
Medical personnel present.
Secondary means of communication.
(2) MRE's per student.

UNITED STATES MARINE CORPS
Marine Corps Detachment
Fort Leonard Wood, MO 65473

1.	LESSON DESIGNATOR:	Interservice Nonlethal Individual Weapons Instructor Course (INWIC). 13A (INIWIC.13A)
2.	LESSON TITLE:	Scenario Practical Exercise
3.	SCHOOL/COURSE:	Nonlethal Individual Weapons Instructor Course
4.	STUDY ASSIGNMENT:	Review course curriculum
5.	ESTIMATED TIME:	8 hours
6.	INSTRUCTORS REQUIRED:	(1) primary, (1) alternate per ratio: 1:10
7.	INSTRUCTOR REFERENCES:	INIWIC POI
8.	DATE PREPARED:	September 2001
9.	PRESENTATION METHOD:	Practical Exercise
10.	TRAINING AIDS:	Warning Order and five paragraph operations order for exercise, rubber rifles/shotguns, batons, shields, Inert MK4 OC canister, bullhorn, radio
11.	STUDENT EQUIPMENT:	(1) 782 gear (1) Kevlar with face shield (1) Shin guards

5. Unit Training Schedule Number 2

Table D-1. NLW Training Course 5-Day Training Schedule

Day/date	*Time*	*Event*	*Location*	*Instructor*	*Notes*
Mon/07 Jan 02	0730-0745	Welcome Aboard	Bldg RR202	MSgt Dixon	
T-1	0745-0845	Force Continuum	SOTG	TBD	L
	0900-1015	MOAB	SOTG	TBD	L
	1030-1130	OC	SOTG	TBD	L
	1130-1300	Noon Meal	Mess Hall	Company GySgt	
	1300-1400	OC Practical Application	SOTG	TBD	PA
	1400-1700	OC Certification	SOTG	TBD	PA
Tues/08 Jan 02	0730-0830	ROE	SOTG	TBD	L
T-2	0845-0945	Crowd Control	SOTG	TBD	L
	1000-1130	MDTS	SOTG	TBD	L, PA
	1130-1300	Noon Meal	Mess Hall	Company GySgt	
	1300-1700	MDTS	SOTG	TBD	L, PA
Wed/09 Jan 02	0730-1130	Batons	SOTG	TBD	L, PA
T-3	1130-1300	Noon Meal	Mess Hall	Company GySgt	
	1300-1500	Batons	SOTG	TBD	L, PA
	1500-1700	Formations	SOTG	TBD	L
Thurs/10 Jan 02	0730-1130	Formations	SOTG	TBD	PA
T-4	1130-1300	Noon Meal	Mess Hall	Company GySgt	
	1300-1700	Nonlethal Munitions/shotguns/ M2O3/sustainment training	Square bay/M203	TBD	L, PA
Fri/11 Jan 02	0730-1130	Nonlethal Munitions/stinger ball grenades/shotgun/M2 03	Square Bay/M203	TBD	PA
T-5	1130-1300	Noon Meal	Mess Hall	Company GySgt	
	1300-1700	Nonlethal Munitions/stinger ball grenades/shotgun/M2 03	Nonlethal Range 1&2	TBD	PA

Notes:

L = Lecture
PA = Practical Application

MDTS = Manadnock Defensive Tactics System
MOAB = Management of Aggressive Behavior
SOTG = Special Operations Training Group (USMC training unit)

6. Sample Unit Training Schedule Number 3

Table D-2. NLW Training Course 10-Day Training Schedule

DAY/DATE	TIME	EVENT	LOCATION	INSTRUCTOR	NOTES
Mon/29 Oct 01 T-1	0730-0745	Welcome Aboard	Bldg RR-202	MSgt Dixon	
	0745-0845	Force Continuum	PT Field	Sgt Whitworth	L
	0900-1015	MOAB	PT Field	GySgt German	L
	1030-1130	OC	PT Field	Sgt Hoskins	L
	1130-1300	Noon Meal	Mess Hall	Company Gunny	
	1300-1400	OC Practical Application	PT Field	Sgt Hoskins	PA
	1400-1700	OC Certification	PT Field	Sgt Hoskins	PA
Tues/30 Oct 01 T-2	0730-0830	ROE	PT Field	Cpl Fewox	L
	0845-0945	Crowd Control	PT Field	SSgt Parrish	L
	1000-1130	MDTS	PT Field	Sgt Johnson	L, PA
	1130-1300	Noon Meal	Mess Hall	Company Gunny	
	1300-1700	MDTS	PT Field	Sgt Johnson	L, PA
Wed/31 Oct 01 T-3	0730-1130	Batons	PT Field	Sgt Johnson	L, PA
	1130-1300	Noon Meal	Mess Hall	Company Gunny	
	1300-1500	Batons	PT Field	Sgt Hoskins	L, PA
	1500-1700	Formations	PT Field	GySgt German	L
Thur/1 Nov 01 T-4	0730-1130	Formations	PT Field	Sgt Whitworth	PA
	1130-1300	Noon Meal	Mess Hall	Company Gunny	
	1300-1700	Nonlethal Munitions	Nonlethal Range 1&2	GySgt German	L, PA
Fri/2 Nov 01 T-5	0730-1130	Nonlethal Munitions	Nonlethal Range 1&2	SSgt Parrish	PA
	1130-1300	Noon Meal	Mess Hall	Company Gunny	
	1300-1700	Nonlethal Munitions	Nonlethal Range 1&2	SSgt Parrish	PA
Mon/5 Nov 01 T-6	0730-1130	Nonlethal Grenades Formation Fire	Nonlethal Range 1&2	GySgt German	L, PA
	1130-1300	Noon Meal	Mess Hall	Company Gunny	
	1130-1700	Formation Fire/Sustainment Training	Nonlethal Range 1&2	SSgt Blackler	PA
Tues/6 Nov 01 T-7	0730-0830	Introduction to MOOTW	PT Field	Maj Desy	L
	0830-1130	Vehicle Checkpoints/Vehicle Searches	PT Field	Sgt Whitworth	L, PA
	1130-1300	Noon Meal	Mess Hall	Company Gunny	
	1300-1500	Vehicle	PT Field	SSgt Parrish	PA

Table D-2. NLW Training Course 10-Day Training Schedule

DAY/DATE	TIME	EVENT	LOCATION	INSTRUCTOR	NOTES
		Checkpoints/Vehicle Searches			
	1500-1700	Roadblocks/Personnel Searches	PT Field	SSgt Blackler	L, PA
Wed/7 Nov 01	0730-1130	Roadblock/Personnel Searches	PT Field	SSgt Blackler	PA
T-8	1130-1300	Noon Meal	Mess Hall	Company Gunny	
	1300	Operations Order	Nonlethal Classroom	MSgt Dixon	Company Staff
Thurs/8 Nov 01	0730-1000	Confirmation Brief	PT Field	Maj Desy	
T-9	1000-1200	Insert	Combat Town	SSgt Parrish	
	1200-0000	STX	Combat Town	SSgt Parrish	
Fri/9 Nov 01	0001-1200	STX	Combat Town	SSgt Parrish	
T-10	1200-1300	Cleanup/Debrief	Combat Town	Company Gunny	
	1300	Graduation	Combat Town	SOTG OIC	

Notes:
L = Lecture
PA = Practical Application
MOAB = Management of Aggressive Behavior
MDTS = Manadnock Defensive Tactics System
STX = Situational Training Exercise

Appendix E

NLW Capability Sets

This appendix provides the current NLW capability sets for the services. These sets continue to be upgraded and changed according to service needs and the availability of new technologies. The U.S. Army and USMC capability sets are standard for the service, while the USAF, USN, and USCG are still developing and fielding their sets.

1. U.S. Army NLW Capability Set

Table E-1 U.S. Army Nonlethal Capability Set (Version FY 2001)

	Name	*Quantity*	*Unit Cost*	*Item Total*
GSAPMDK5X250AFHC	Non-Ballistic Riot Face Shields	200	$42.50	$8,500.00
GSAA TIPMBS9	Non-Ballistic Riot Body Shields Paulson	150	$88.50	$13,275.00
GSAATIUSMCSG	Non-Ballistic Riot Shin Guards	150	$60.00	$9,000.00
PT702MTUSA	Ballistic Face Shields	10	$246.00	$2,460.00
PT2237	Ballistic Body Shields	10	$1,052.00	$10,520.00
GSASF6SL	Sure Fire Field Light	10	$185.82	$1,858.20
MON2800 w/BCS-1	36" Hickory Riot Baton with Grommet	200	$15.50	$3,100.00
EADBH	Eagle Duty Baton Holder OD	200	$5.50	$1,100.00
GSAATIA2M	Portable Bullhorn	10	$82.00	$820.00
AAPB35W1	Ground Mount Bullhorn	8	$815.00	$6,520.00
AAPBWH10001	Megavox Wireless MIC	8	$0.00	$0.00
AASS450	Megavox Stand	8	$0.00	$0.00
GSASF6P	Individual High Intensity Light	15	$42.74	$641.10
GSAATIEATAVLFP6POD	Eagle Sure Fire 6P Holster	15	$8.75	$131.25
GSASFP60	Individual High Intensity Light Accessory Kit	15	$12.53	$187.95
XENNCSUSA	High Intensity Light	8	$2,795.00	$22,360.00
NIKFLEXCUFF	Disposable Forearm/Ankle Cuffs (Single)	1000	$0.70	$700.00
NIKCUTTER	Flex Cuff Cutters	2	$9.00	$18.00
GSAATIEAMK4	MK-4 Pouch OD	200	$6.75	$1,350.00
GSADEF5149	Inert Individual Training (OC) Dispenser	200	$6.07	$1,214.00
GSASF622	12-Gauge Shotgun High Intensity Light	50	$139.65	$6,982.50
GSAATIEAUPMS	12-Gauge Utility (25-Round) Pouch	50	$23.65	$1,182.50
GSAATISSSCQTR	12-Gauge Duttcuff Carrier (6-Round)	50	$10.75	$537.50
GSAATISASDD60D	Diversionary/Sting Ball Grenade Pouch	25	$47.25	$1,181.25
GSAATIEA40MM6	40-mm Carrying Pouch	100	$13.65	$$1,365.00
GSAATIRSC3	Caltrop	250	$10.50	$2,625.00
GSAATIFS333	Riot Training Suit with Accessories	4	$638.50	$2,554.00
GSAMONUTB	Riot Training Bag	10	$80.00	$800.00
BNCCELLS	C-Cell Batteries	0	$1.00	$0.00
BNAAACELLS	AA Batteries	0	$0.75	$0.00
GSASFDL123A	3V Lithium Batteries	0	$2.85	$0.00
BN9V	9V Batteries	0	$2.00	$0.00
ATIPMI	Packaging Marking and Integration Fee			$20,300.00
	Hardware Tools		**Item Total**	**$125,683.25**
	U.S. Army NLW Capability Set	**5**	**$125,683.25**	**$628,416.25**

2. USMC Nonlethal Weapons Capability Set

a. USMC NLW Company-Size Set

Table E-2 U.S. Marine Corps FY02 Company-Size Nonlethal Capability Set

TAMCN C41202E	*NSN 9999-01-464-2285*	*SET A (Company Size)*		
NSN/Product ID	***Name***	***Quantity***	***Unit cost***	***Item Total***
GSAATINCSPMDK5	Non-Ballistic USMC Riot Face Shields (S/R)	200	$42.50	$8,500.00
GSAATIPMBS9	Non-Ballistic Riot Body Shields Paulson	60	$88.50	$5,310.00
GSAATIUSMCSG	Non-Ballistic Riot Shin Guards	200	$60.00	$12,000.00
GSAATI23/36	USMC 23"/36" Riot Baton (Expandable) (w/Holster)	200	$73.25	$14,650.00
GSAATITRAINING BATON	USMC Riot Baton (Practice) (Price TBD)	200	$20.00	$4,000.00
GSAATIA2M	Portable Bullhorn	12	$82.00	$984.00
GSAATIDC/W	USMC Disposable Handcuffs	400	$0.94	$376.00
GSAATIDC/T	USMC Disposable Handcuff Training Device	100	$0.94	$94.00
GSAATIEAMK4	MK-4 Pouch OD	200	$6.75	$1,350.00
GSADEF5049	MK-4 Individual Live (OC) Dispenser (Pepper Spray)	200	$7.29	$1,458.00
GSADEF5099	MK-9 Live Squad (OC) Dispenser (Pepper Spray)	40	$26.20	$1,048.00
GSATIEAMK9	MK-9 Pouch OD	40	$11.00	$440.00
GSAATIDEF5846L	MK-46 Live (OC) (Horizontal) (Pepper Spray)	12	$300.00	$3,600.00
GSAATIDEF1515	MK-46 Sling	12	$30.25	$363.00
GSAATIDEF5520	MK-46 Hardware Kit	4	$68.50	$274.00
GSAATIDEF5546	MK-46 Refill Unit (OC) Live (Pepper Spray)	4	$46.25	$185.00
GSADEF5149	MK-4 Inert Individual Training (OC) Dispenser	200	$6.07	$1,214.00
GSADEF5199	MK-9 Inert Squad Training (OC) Dispenser	40	$20.10	$804.00
GSAATIDEF5946L	MK-46 Inert Training (OC) (Horizontal) (w/Dye)	12	$240.00	$2,880.00
GSADEF5546I	MK-46 Refill Unit (OC) Inert Training (w/Dye)	4	$27.50	$110.00
GSAATIFS333	Riot Training Suit w/Accessories (F.I.S.T.)	12	$638.50	$7,662.00
GSAMONUTB	Riot Training Strike Bag	24	$80.00	$1,920.00
GSAATIRSC3	Caltrop	400	$10.50	$4,200.00
TBD	Stinger Road Side Spike Kit (15.5') (Price TBD)	12		
XENNLCSUSMC	High Intensity Light (Xenonics)	6	$2,975.00	$17,850.00
GSAATIEATAVLFP6POD	Eagle Sure Fire 6P Holster	60	$8.75	$525.00
GSASF6P	Individual High Intensity Light	60	$42.74	$2,564.40
GSASFP60	Individual High Intensity Light Accessory Kit	30	$12.53	$375.90
GSASFDL123A	3V Lithium Batteries	120	$2.85	$342.00
GSAATISSSCQTR	12-Gauge Buttstock Cuff	36	$10.75	$387.00
GSAATIEAUPMS	12-Gauge Utility (25-Round) Pouch	72	$23.65	$1,702.80
GSAATIEA40MM6	40-mm Carrying Pouch	36	$13.65	$491.40
GSAATISASDD60D	Diversionary/Sting Ball Grenade Pouch	60	$47.25	$2,835.00
CSI9512	12-Gauge Launching Cup (Availability and Price TBD)	36		
1005-01-371-5639/ 1005-01-251-8578	12-Gauge Mossberg 500A2 / 590 Shotgun	36	$250.00	$9,000.00
ATIPMI	Packaging Marking and Integration Fee			$11,578.00
	Hardware Totals		**Item Total**	**$121,073.50**

b. USMC NLW Platoon-Size Set

Table E-3 USMC FY02 NLW Capability Set—Platoon Size

TAMCN C41202E	*NSN 9999-01-464-2285*	*SET B (Platoon Size)*		
NSN/Product ID	***NAME***	***Quantity***	***Unit Cost***	***Item Total***
GSAATINCSPMDK5	**Non-Ballistic USMC Riot Face Shields (S/R)	50	$42.50	$2,125.00
GSAATIPMBS9	Non-Ballistic Riot Body Shields Paulson	15	$88.50	$1,327.50
GSAATIUSMCSG	Non-Ballistic Riot Shin Guards	50	$60.00	$3,000.00
GSAATI23/36	USMC 23"/36" Riot Baton (Expandable) (w/Holster)	50	$73.25	$3,662.50
GSAATITRAINING BATON	USMC Riot Baton (Practice) (Price TBD)	50	$20.00	$1,000.00
GSAATIA2M	Portable Bullhorn	3	$82.00	$246.00
GSAATIDC/W	USMC Disposable Handcuffs	100	$0.94	$94.00
GSAATIDC/T	USMC Disposable Handcuff Training Device	25	$0.94	$23.50
GSAATIEAMK4	MK-4 Pouch OD	50	$6.75	$337.50
GSADEF5049	MK-4 Individual Live (OC) Dispenser (Pepper Spray)	50	$7.29	$364.50
GSADEF5099	MK-9 Live Squad (OC) Dispenser (Pepper Spray)	10	$26.20	$262.00
GSATIEAMK9	MK-9 Pouch OD	10	$11.00	$110.00
GSAATIDEF5846L	MK-46 Live (OC) (Horizontal) (Pepper Spray)	3	$300.00	$900.00
GSAATIDEF1515	MK-46 Sling	3	$30.25	$90.75
GSAATIDEF5520	MK-46 Hardware Kit	2	$68.50	$137.00
GSAATIDEF5546	MK-46 Refill Unit (OC) Live (Pepper Spray)	2	$46.25	$92.50
GSADEF5149	MK-4 Inert Individual Training (OC) Dispenser	50	$6.07	$303.50
GSADEF5199	MK-9 Inert Squad Training (OC) Dispenser	10	$20.10	$201.00
GSAATIDEF5946L	MK-46 Inert Training (OC) (Horizontal) (w/Dye)	3	$240.00	$720.00
GSADEF5546I	MK-46 Refill Unit (OC) Inert Training (W/Dye)	2	$27.50	$55.00
GSAATIFS333	Riot Training Suit W/Accessories (F.I.S.T.)	3	$638.50	$1,915.50
GSAMONUTB	Riot Training Strike Bag	3	$80.00	$240.00
GSAATIRSC3	Caltrop	50	$10.50	$525.00
TBD	Stinger Road Side Spike Kit (15.5') (Price TBD)	3		
XENNLCSUSMC	High Intensity Light (Xenonics)	4	$2,975.00	$11,900.00
GSAATIEATAVLFP6POD	Eagle Sure Fire 6P Holster	50	$8.75	$437.50
GSASF6P	Individual High Intensity Light	50	$42.74	$2,137.00
GSASFP60	Individual High Intensity Light Accessory Kit	25	$12.53	$313.25
GSASFDL123A	3V Lithium Batteries	100	$2.85	$285.00
GSAATISSSCQTR	12-Gauge Buttstock Cuff	15	$10.75	$161.25
GSAATIEAUPMS	12-Gauge Utility (25 Round) Pouch	30	$23.65	$709.50
GSAATIEA40MM6	40-mm Carrying Pouch	50	$13.65	$682.50
GSAATISASDD60D	Diversionary/Sting Ball Grenade Pouch	50	$47.25	$2,362.50
CSI9512	12-Gauge Launching Cup (Availability and Price TBD)	15		
1005-01-371-5639/ 1005-01-251-8578	12-Gauge Mossberg 500A2 / 590 Shotgun	15	$250.00	$3,750.00
ATIPMI	Packaging Marking and Integration Fee			$3,000.00
	Hardware Totals		**Item Total**	**$43,471.25**

** Riot shield. Marine Corps security forces are currently evaluating to use in conjunction with each other as listed below. They are not currently deployed but are undergoing field evaluation.

(1) The six feet Armadillo Interlocking Shield for the hard baseline.

(2) The four feet.. Armadillo Interlocking Shield for the soft baseline.

3. USN NLW Capability Set

a. The USN is in the process of developing NLW capabilities and capability sets.* The sets will support the specific areas of operation, mission, and/or user.

b. The first version will address the force protection mission and will be based on other service-approved items and some longer-range commercial-off-the-shelf items currently being upgraded for military use. The mobile security forces, master at arms, and shipboard force protection units are the targeted users.

c. A notional capability set for the Mobile Security Forces is listed below.

Table E-4. Notional USN Capability Set

Name	*Nomenclature (Mfr. Representative)*	*Quantity*	*Unit Cost*	*Case Size*	*Cost Extended*
Fin Stabilize Round		1 Case	$5	100	$400
Stingshot		1 Case	$5	100	$400
Shotbag		1 Case	$5	100	$450
Flamethrower Round		1 Case	$8	100	$750
Tactical Blast Bursting Round		1 Case	$7	100	$650
Ultraflash Bursting Round		1 Case	$8	100	$750
Launchable Stun Munition	POP, Inc. T-444	1 Case	$60	24	$1,440
40-mm Sparkshot		1 Case	$35	24	$840
40-mm Stingshot		1 Case	$25	24	$600
40-mm Shotbag		1 Case	$25	24	$600
40-mm OC		1 Case	$28	24	$660
Stinger	POP, Inc. I-452	1 Case	$60	24	$1,440
Stun	POP, Inc. T-470	1 Case	$55	24	$1,320
Flash/Bang Multi	POP, Inc. D-451	1 Case	$65	24	$1,560
OC	POP, Inc. I-452	1 Case	$65	24	$1,560
12-Gauge Launching Blanks		200	$3		$600
Grenade Training Rounds		12	$65		$780
12-Gauge Training Rounds		24	$3		$72
40-mm Training Rounds		12	$20		$240
12-Gauge Shotguns	Standard Issue/Modified	12	$1,000		$12,000
12-Gauge Launching Adapter		6	$325		$1,950
Face Shields		15	$100		$1,500
Helmets		15	$200		$3,000

* In FY 01, the USN fielded an USMC capability set to Naval Air Station, Roosevelt Roads, Puerto Rico.

Table E-4. Notional USN Capability Set

Name	*Nomenclature (Mfr. Representative)*	*Quantity*	*Unit Cost*	*Case Size*	*Cost Extended*
Tactical Vest		15	$300		$4,500
Bullhorn		2	$100		$200
Medical Kit		2	$50		$100
Searchlights		2	$100		$200
Field Pack		15	$100		$1,500
Holsters/Belts		15	$50		$750
Ordnance Packaging, Handling, Storage and Transportation (PHS&T)					$1,500
Misc. (Certification, Training and Documentation)					$3,000
			Total Cost Basic Kit		**$45,312**

Table E-5. Notional USN Capability Set

Name	*Nomenclature (Mfr. Representative)*	*Quantity*	*Unit Cost*	*Case Size*	*Cost Extended*
Stingmore Mine (Std Body) with Rubber Ball Projectiles	POP, Inc. I-454	24	$350		$8,400
Stingmore Mine Stand		6	$100		$600
Electric Firing Boxes		8	$35		$280
Area Lighting					$1,000
Electric Generator					$500
Signs and Placards					$500
Personal NVGs and Opticals		6	$2,000		$12,000
Additional PHS&T					$2,000
Misc. (Certification, Training and Documentation)					$2,000
Total Cost Augmented Basic Kit					**$27,280**
Total Cost (Basic plus Augment)					$72,592
Portable Vehicle Arresting Barrier	U.S. Army	1	$40,000		$40,000
Running Gear Entanglement System (RGES)	NSWCDD	1000 ft system	$25,000		$25,000
Antimaterial Round (USCG)		24	$12		$288
Spike Strips		2	TBD		
Improved Sensor (s)					
RF			TBD		

Table E-5. Notional USN Capability Set

Name	*Nomenclature (Mfr. Representative)*	*Quantity*	*Unit Cost*	*Case Size*	*Cost Extended*
IR			TBD		
Acoustic			TBD		
Chemical Detection	SABRE 2000	1	$20,000		$20,000
Ordnance Packaging, Handling, Storage and Transportation (PHS&T)					$6,000
Misc. (Certification, Training and Documentation)					$2,000
Total Cost Vehicle Supplement					**$93,288**
Total Cost Basic, Augmented, and Vehicle Supplement					**$165,880**

4. USAF NLW Capability Set

Table E-6. USAF NLW Capability Set

NSN	*Item Description*	*EA*	*Item $*	*Item Total*
GSADEF5149	Inert Individual (OC) Training Dispenser	1	$6.07	$6.07
GSADEF5846H	Mid-Size Riot Control Dispenser (Chemical)	3	$318.75	$956.25
GSADEF5546	Riot Control Dispenser (OC) Agent Quarts	1	$46.25	$46.25
	12-Gauge Shotgun (Area CDC/Rubber Pellet)	36	$4.46	$160.56
	12-Gauge Shotgun (Point/Fin Stabilized)	36	$2.85	$102.60
CSI (USMC) GG04	Stun Hand (Sting Ball) Grenade	50	$35.00	$1,750.00
M84	Stun Hand Grenade/Flash-Bang Device	50	$78.00	$3,900.00
GSAPMDK5X/AF	Non-Ballistic Face Shields	13	$29.50	$383.50
GSAPMB80	Non Ballistic Full Length Riot Shield	13	$88.50	$1,150.50
GSAUSMCSG	Non-Ballistic Shin guards	13	$6.000	$780.00
GSAATI23/36	36" Riot Baton w/Holster	13	$73.25	$952.25
GSAA2	Portable Megaphone	1	$82.00	$82.00
EAAFCUSTOM	Tactical Vest Rough Estimate	3	$150.00	$450.00
HARG700	Riot Control Gloves	13	$52.20	$678.60
Champro	Protective Cups	13	$7.25	$94.25
	40-mm (Point) Sponge Round	54	$23.00	$1,242.00
	40-mm (Area) Crowd Dispersal Cartridge Grenade	54	$22.00	$1,188.00
GSAF333	Riot Baton Training Suit	1	$638.50	$638.50

Table E-6. USAF NLW Capability Set

NSN	*Item Description*	*EA*	*Item $*	*Item Total*
MONDC/W	Disposable Forearm/Ankle Cuffs (FLEX)	40	$0.94	$37.60
GSADEF5049/ EA	Individual (OC) Dispenser with Carrier	13	$7.29	$94.77
	12-Gauge Utility Pouch	2	$24.00	$48.00
	12-Gauge Buttstock Cuff	3	$11.00	$33.00
	40-mm Carrying Pouch	2	$14.00	$28.00
	Training Strike Bag	2	$80.00	$160.00
	Cost of Clip-Lok Boxes			$480.00
	Approximate freight shipped via commercial freight into CONUS			$1,500.00
			Total Estimate	**$16,942.70**
Proposal to add TASERS to USAF Capability Set after testing.				

5. Notional USCG NLW Capability Set

Table E-7. Notional USCG NLW Capability Set

Reference: (a) COMDTINST M16247.1B Maritime Law Enforcement Manual (MLEM) (b) Project Ericson Website	
Expandable Baton	
Restraint Devices	Hand Cuffs, Flex Cuffs, Leg Irons, Migrant Restraint Harness
Individual OC Spray	
Non-Ballistic Riot Face Shield	
Non-Ballistic Riot Body Shield	
Non-Ballistic Riot Shin Guard	
Running Gear Entanglement Systems	

The following information describes each of the components of the NLW capability set currently deployed by the USCG. (The USCG has not designated a standard set of nonlethal gear. The using unit determines the gear needs.)

6. Restraint Devices

a. Handcuffs are the preferred restraint devices when dealing with a small number of subjects. Handcuffs should be steel, chain-linked construction capable of being double locked, with nickel, blued, or stainless steel finish.

b. Using flexible handcuffs for the temporary restraint of subjects is also authorized. Since they do not require a key and are relatively easy to apply, flexible handcuffs are particularly useful and appropriate when dealing with large numbers of subjects (e.g. migrants).

c. Leg irons are authorized restraint devices and may be a preferred option for restraining subjects for extended periods of time.

d. Migrant restraint harnesses are authorized when dealing with large numbers of subjects.

7. Expandable Baton

The expandable baton is the only intermediate weapon authorized for USCG personnel to carry. It should have a steel shaft that is silver in color and corrosion resistant (e.g. chrome, nickel, stainless steel) and have no knurling or sharp edges. It should have a foam or similar, non-slip grip. It should be rigid when deployed and must be no less than 15 inches and no greater than 22 inches when expanded. Spring loaded or other trigger style batons are not authorized.

8. OC Pepper Spray

Only chemical irritants containing the active ingredient OC are authorized for USCG personnel to carry. The aerosol must contain no greater than 10 percent and no less than 5 percent OC concentration. It shall have a water-based carrier, be nonflammable, nontoxic and produce a straight stream delivery (not fog).

9. Helmet Riot Face Shield

Comparable to DOD type. Not currently deployed. Undergoing field evaluation.

10. Riot Shield

Two shields are currently being evaluated. A 24 x 48 inches clear lexan weighing 14 pounds and a smaller version measuring 20 x 36 inches. They are not currently deployed. Undergoing field evaluation.

REFERENCES

Joint

JP 1-02, *DOD Dictionary of Military and Associated Terms*, 12 Apr. 2001

JP 3-07, *Joint Doctrine for Military Operations Other Than War*, 16 Jun. 1995

JP 3-07.3, *JTTP for Peace Operations*, 12 Feb.1999

JP 3-57, *Doctrine for Joint Civil-Military Operations*, 8 Feb. 2001

JP 3-58, *Joint Doctrine for Military Deception*, 31 May 1996

JP 3-61, *Doctrine for Public Affairs in Joint Operations*, 14 May 1997

JP 4-0, *Joint Doctrine for Logistic Support of Joint Operations*, 6 Apr. 2000

Multi-Service

FM 3-100.12, MCRP 5-12.1C, NTTP 5-03.5, AFTTP(I) 3-2.34, *Risk Management* for *Multiservice Tactics, Techniques, and Procedures*, 15 Feb. 2001

FM 3-31.1, MCWP 3-36, *Army and Marine Corps Integration in Joint Operations*, 21 Nov. 2001

Army

AR 190-14, *Military Police Carrying of Firearms and Use of Force for Law Enforcement and Security Duties*, 12 Mar. 1993

AR 350-38, *Training Devices Policies and Management*, 15 Oct. 1993

AR 27-1, *Judge Advocate Legal Services*, 30 Sep. 1996

FM 3-0, *Operations*, 14 Jun. 2001

FM 3-05.30 (33-1), *Psychological Operations*, 19 Jun. 2000

FM 19-15, *Civil Disturbances*, 25 Nov. 1985

FM 27-10, *The Law of Land Warfare*, 18 Jul. 1956

FM 100-19, *Domestic Support Operations*, 1 Jul. 1993

TC 90-1, *Training for Urban Operations*, 1 Apr. 2002

TRADOC Pamphlet 525-5, *Force XXI Operations, A Concept for the Evolution of Full Dimensional Operations for the Strategic Army of the Twenty-First Century*, 1 Aug. 1994

TRADOC Pam 350-9, *TRADOC Training Devices for Army wide Use*, 30 Sep. 1993

TRADOC Regulation 350-70, *Systems Approach to Training Management, Processes, and Products*, 9 Mar. 1999.

TRADOC Regulation 385-2, chg.1, TRADOC Safety Program, 10 Oct. 2000.

Marine Corps

MCWP 3-33.4, *Domestic Support Operations*, Jul. 1993

I MEF FMF, After Action Report—UNITED SHIELD, 7 Jun. 1995

MCO 1510.112, *Individual Training Standards for Nonlethal Weapons*, 27 May 1998

MCO 5500.6F, *Arming of Security and Law Enforcement Personnel and the Use of Force*, 20 Jul. 1995

MCRP 5-12.1A, *The Law of Land Warfare.*

MCO 1630.4A, *Law Enforcement and Physical Security Activities*, 19 Jun. 1991.

MCO 1510.114, *Individual Training Standards (ITS) for Antiterrorism Force Protection (AT/FT) System*, 30 Jan. 1998.

MCWP 3-11.1, *Marine Rifle Company*, Feb. 1978

MCRP 3-11.1A, *Commander's Tactical Handbook*, 30 Jul. 1988. .

MCWP 3-11.2 *Marine Rifle Squad*

Marine Corps Gazette, *"Less-Lethal" Force in Operation UNITED SHIELD*, Sep. 1995.

Navy

Naval Doctrine Publication 1, Naval Warfare (U), 28 Mar. 1994

NWP 1-14M, *The Commander's Handbook on the Law of Naval Operations*, Oct. 1995

SECNAVINST 5500.29B, *Use of Deadly Force and the Carrying of Firearms by Personnel of the Department of the Navy in Conjunction with Law Enforcement Security Duties and Personal Protection*, 29 Sep. 1992

Air Force

AFI 31-201, *Security Police Standards and Procedures*, 4 Dec. 2001

AFI 31-207, *Arming and Use of Force by Air Force Personnel*, 1 Sep. 1999

AFMAN 31-201V6, *Civil Disturbance*, 17 May 2002

AFMAN 31-222, *Security Forces Use of Force Manual*, 1 Jun. 2001

Coast Guard

COMDTINST M16247.1B, Maritime Law Enforcement Manual, 9 Aug. 2001

Other

DODD 3000.3-R, *Policy for Nonlethal Weapons*, 9 Jul. 1996

DODD 3025.1, *Military Support to Civil Authorities (MSCA)*, 15 Jan. 1993

DODD 5100.77, *DOD Law of War Program*, 9 Dec. 1998

DODD 3025, *I-M Manual for Civil Emergencies*, 2 Jun. 1994

DODD 3025.12, *Military Assistance for Civil Disturbances*, 4 Feb. 1994

DODD 3025.13, *DOD Resources in Support of US Secret Service*, 13 Sep. 1985

DODI 5000.2, *Operation of the Defense Acquisition System*, 5 Apr. 2002

DODR 5000.1, *Defense Acquisition*, 23 Oct. 2000, Ch 1, 4 Jan. 2001

DOD Joint Concept for Nonlethal Weapons, 5 Jan. 1998

DODD 3025.12, *Military Assistance for Civil Authorities*, 18 Feb. 1997

DODD 3025.16, *Military Assistance for Civil Authorities Liaison Office (EPLO) Program,*18 Dec. 2000

DODI 5000.2, *Operation of the Defense Acquisition System*, 5 Apr. 2002

Mission Need Statement (MNS), *Range of Nonlethal Force Capabilities (NLFC*), 10 Oct. 1996

A Guide for Effective Peacetime Employment of Reserve Components Units and Individuals, OASD RA (M&P), Dec. 2000

Defense and Intelligence Abbreviations and Acronyms, Nov. 1997

Emergency Management Assistance Compact—Guide Book and Operating Procedures, National Emergency Management Association, Aug. 2000

Inside the Navy, Marine Corps Nonlethal Weapons Experiment Underway at Camp Pendleton, CA, Vol. 9, No. 29, Jul. 22, 1996

Military Support To Civil Authorities—Interagency Disaster Resource Guide, National Interagency Civil-Military Institute, Jan. 2001

National Guard Bureau Military Support to Civil Authorities Lessons Learned, 6 Sep. 2000

The Robert T. Stafford Disaster Relief and Emergency Assistance Act, as amended 42 USC 1521, Jul.2001

Joint Universal Lessons Learned System, Operation UNITED SHIELD

Nonlethal Weapons Joint Mission Area Analysis/Joint Mission Needs Analysis, Dec. 2000

Interservice Nonlethal Individual Weapons Instructor Course (INIWIC) Program of Instruction, Oct.2000

Presidential Decision Directive PDD-39 U.S. Policy on Counterterrorism 21 Jun.1995

PDD-62 Combating Terrorism—Fact Sheet, 22 May 1998

The Federal Response Plan Notice of Change 7, Feb.1997; FEMA, Number 229 Change 11

Federal Response Plan, Apr 1999

GLOSSARY

PART I – ABBREVIATIONS AND ACRONYMS

A

ACOG	advanced combat optical gunsight
ADS	aerial delivery system
AFDC	Air Force Doctrine Center
AFI	Air Force instruction
AFMAN	Air Force manual
ALSA	Air Land Sea Application
AMC	Army Materiel Command
AO	area of operation
AR	Army regulation
ATSC	Army Training Support Center

B

bldg	building
BLT	battalion landing team
bn	battalion

C

C4ISR	command, control, communications, computers, intelligence, surveillance, and reconnaissance
CA	civil affairs
cal	caliber
CANS	compressed air net system
CDC	crowd disposal cartridge
CMC	Commandant of the Marine Corps
COB	close of business
COC	combat operations center
COMDTPUB	Commandant, Coast Guard Publication
COMDTINST	Commandant, Coast Guard Instruction
CONUS	continental United States
COTS	commercial off-the-shelf
CP	command post
CQB	close quarters battle
CS	O-Chlorobenzyl-Malononitrile (riot control agent (tear gas))
CSSOC	combat service support operations center
cu ft	cubic feet

D

DA	Department of the Army
dB	decibel
DECON	decontamination
demo	demonstration
DM	designated marksman
DOD	Department of Defense
DODD	Department of Defense Directive
DODI	Department of Defense Instruction
DODIC	Department of Defense Identification Code
DODAC	Department of Defense Ammunition Code
DOT	Department of Transportation
DP	downed person
DPB	deployable pursuit boat
DTD	dated

E

EADBH	Eagle Duty Baton Holder
EEZ	exclusive economic zone
EOD	explosive ordnance disposal

F

FAST	Fleet Antiterrorism Security Team
FDP	food distribution point
FLEX	forearm/ankle cuff
FM	field manual
four M's	mind, mass, movement, muscle
FSE	fire support element
ft	feet

G

G1	Army or Marine Corps component manpower or personnel staff officer (Army division or higher staff, Marine Corps brigade or higher staff)
G2	Army or Marine Corps component intelligence staff officer (Army division or higher staff, Marine Corps brigade or higher staff)
G3	Army or Marine Corps component operations staff officer (Army division or higher staff, Marine Corps brigade or higher staff)

G4	Army or Marine Corps component logistics staff officer (Army division or higher staff, Marine Corps brigade or higher staff)
ga	gauge
GOTS	government off-the-shelf
H	
HMMWV	high mobility multipurpose wheeled vehicle
I	
IADS	Integrated Air Defense System
IAW	in accordance with
ID	identify/identification
IFF	identification, friend or foe
in.	inch
INIWIC	Interservice Nonlethal Individual Weapons Instructor Course
IPB	intelligence preparation of the battlespace
ITRO	Interservice Training Organization
J	
JMAA	Joint Mission Area Analysis
JTTP	joint tactics, techniques, and procedures
K	
KFOR	Kosovo forces
L	
LBE	load-bearing equipment
LOAC	Law of Armed Conflict
LP	listening post
LVOSS	light visual optical screening system
M	
M16	M16A2 rifle
M203	M203 Grenade Launcher
MARCORSYCOM	Marine Corps System Command
MCCDC	Marine Corps Combat Development Command
MCCM-GE	modified crowd control munition—ground emplaced
MCO	Marine Corps Order
MCPDS	Marine Corps Doctrinal Publication System
MCRP	Marine Corps reference publication

MCSF	Marine Corps security force
MCWP	Marine Corps warfighting publication
MDST	Monadnock Defensive Tactics System
MEDEVAC	medical evacuation
MEF	Marine expeditionary force
METT-T	mission, enemy, terrain and weather, troops and support available – time available
METT-TC	mission, enemy, terrain and weather, troops and support available – time available and civil considerations
MEU	Marine expeditionary unit
MIO	Maritime interception operation
MLEM	Maritime Law Enforcement Manual
MILSTRIP	military
mm	millimeter
MNS	mission need statement
MOAB	management of aggressive behavior
MOOTW	military operations other than war
MOS	military occupational specialty
MOUT	military operations in urban terrain
MP	military police
mph	miles per hour
MRE	meal, ready to eat
MTTP	multi-Service tactics, techniques, and procedures

N

NATO	North Atlantic Treaty Organization
NAVSOP	Navy standing operating procedures
NDC	Naval Doctrine Command
NLFC	nonlethal force capabilities
NLW	nonlethal weapon
NSN	national stock number
NSWC	Naval Surface Warfare Command
NWDC	Naval Warfare Development Command
NWP	Naval Warfare Publication

O

OC	Oleoresin Capsicum
OCAT	Oleoresin Capsicum aerosol training
OCONUS	outside the continental United States
OD	olive drab
OP	observation post

OPFOR	opposing force

P

PA	public affairs
PAO	public affairs officer
PASGT	personnel armor system, ground troops
POI	program of instruction
PSYOP	psychological operations
PT	physical training
PVAB	portable vehicle arresting barrier
PVC	polyvinyl chloride

Q

QRF	quick reaction force

R

R&D	research and development
RCA	riot control agent
RGES	running gear entanglement system
ROE	rules of engagement
RRF	rapid reaction force

S

S1	battalion or brigade manpower or personnel staff officer (Army; Marine Corps battalion or regiment)
S2	battalion or brigade intelligence staff officer (Army; Marine Corps battalion or regiment)
S3	battalion or brigade operations staff officer (Army, Marine Corps battalion or regiment)
S4	battalion or brigade logistics staff officer (Army; Marine Corps battalion or regiment)
SASO	stability and support operations
SB	shield bearer
sec	second
SECNAVINST	Secretary of the Navy instruction
SG	shotgun
SJA	staff judge advocate
SNCO	staff noncommissioned officer
SOP	standard operating procedure
SOTG	Special Operations Training Group
SROE	standing rules of engagement
SSC	small-scale contingency

T

TC	training circular
TF	task force
T/O	table of organization
TOC	tactical operations center
TRADOC	United States Army Training and Doctrine Command
TTP	tactics, techniques, and procedures

U

UN	United Nations
US	United States
USA	United States Army
USAF	United States Air Force
USAMPS	United States Army Military Police School
USCG	United States Coast Guard
USMC	United States Marine Corps
USN	United States Navy
USS	United States Ship

V

VCR	videocassette recorder
VLNL	vehicle launched nonlethal grenade
VMNLGL	vehicle mounted nonlethal grenade launcher

W

w/	with
WMD	weapons of mass destruction

PART II – TERMS AND DEFINITIONS

Air Force specialty code (AFSC). Primary job of a U.S. Air Force officer or enlisted person. Analogous to the Army and Marine Corps military occupational skill (MOS).

area of operation (AO). An operational area defined by the joint force commander for land and naval forces. Areas of operation do not typically encompass the entire operational area of the joint force commander, but should be large enough for component commanders to accomplish their missions and protect their forces. (JP 1-02)

battalion landing team (BLT). In an amphibious operation, an infantry battalion normally reinforced by necessary combat and service elements; the basic unit for planning an assault landing. (JP 1-02)

civil affairs. Designate active and Reserve Component forces and units organized, trained, and equipped specifically to conduct civil affairs activities and to support civil-military operations. (JP 1-02)

civil-military operations. The activities of a commander that establish, maintain, influence, or exploit relations between military forces, governmental and nongovernmental civilian organizations and authorities, and the civilian populace in a friendly, neutral, or hostile operational area in order to facilitate military operations, to consolidate and achieve operational U.S. objectives. Civil-military operations may include performance by military forces of activities and functions normally the responsibility of the local, regional, or national government. These activities may occur prior to, during, or subsequent to other military actions. They may also occur, if directed, in the absence of other military operations. Civil-military operations may be performed by designated civil affairs, by other military forces, or by a combination of civil affairs and other forces. (JP 1-02)

combat service support. The essential capabilities, functions, activities, and tasks necessary to sustain all elements of operating forces in theater at all levels of war. Within the national and theater logistics systems, it includes, but is not limited to, that support rendered by service forces in ensuring the aspects of supply, maintenance, transportation, health services, and other services required by aviation and ground combat troops to permit those units to accomplish their missions in combat. Combat service support encompasses those activities at all levels of war that produce sustainment to all operating forces on the battlefield. (JP 1-02)

command and control. The exercise of authority and direction by a properly designated commander over assigned and attached forces in the accomplishment of the mission. Command and control functions are performed through an arrangement of personnel, equipment, communications, facilities, and procedures employed by a commander in planning, directing, coordinating, and controlling forces and operations in the accomplishment of the mission. (JP 1-02)

commercial off the shelf (COTS). Describes items procured for military use through civilian contractors.

core capability. The set of functions that define a significant, stand-alone, operationally effective and suitable military capability such that, should no further development occur, the user will have received a significant capability.

designated marksman (DM). A rifleman stationed in an overwatch position above friendly forces, armed with a standard infantry weapon appropriate to his service that is mounted with a high-powered scope and loaded with lethal ammunition. The designated marksman uses his vantage point to identify agitators to forces below, eliminate lethal threats, and monitor the battlefield.

dispersion. A scattered pastern of hits around the mean point of impact of bombs and projectiles dropped or fired under identical conditions. In chemical and biological operations, the dissemination of agents in liquid or aerosol form. (JP 1-02)

fleet Marine force (FMF). A balanced force of combined arms comprising land, air, and service elements of the U.S. Marine Corps. A fleet Marine force is an integral part of a U.S. Fleet and has the status of a type command. (JP 1-02)

force protection. Security program designed to protect service members, civilian employees, family members, facilities, and equipment, in all locations and situations, accomplished through planned and integrated application of combating terrorism, physical security, operations security, and personal protective services, and supported by intelligence, counterintelligence, and other security programs. (JP 1-02)

government off-the-shelf (GOTS). Analogous to COTS, but describing items already procured by the U.S. government and ready for issue to military units.

host nation. A nation that receives the forces and/or supplies of allied nations, coalition partners, and/or NATO organizations to be located on, to operate in, or to transit through its territory. (JP 1-02)

intelligence preparation of the battlespace (IPB). An analytical methodology employed to reduce uncertainties concerning the enemy, environment, and terrain for all types of operations. Intelligence preparation of the battlespace builds an extensive database for each potential area in which a unit may be required to operate. The database is then analyzed in detail to determine the impact of the enemy, environment, and terrain on operations, and present it in graphic form. Intelligence preparation of the battlespace is a continuing process. Also called IPB. (JP 1-02)

interrogator-translator team (ITT). Marine Corps unit, usually working as a small detachment to a larger infantry unit, highly trained in interrogation techniques and fluent in adversary native language.

joint force commander. A general term applied to a combatant commander, subunified commander, or joint task force commander authorized to exercise combatant command (command authority) or operational control over a joint force. (JP 1-02)

logistics. The science of planning and carrying out the movement and maintenance of forces. In its most comprehensive sense, those aspects of military operations which deal with:

a. Design and development, acquisition, storage, movement, distribution, maintenance, evacuation, and disposition of materiel.

b. Movement, evacuation, and hospitalization of personnel.

c. Acquisition of construction, maintenance, operation, and disposition of facilities.

d. Acquisition of furnishing of services. (JP 1-02)

low-intensity conflict. Political-military confrontation between contending states or groups below conventional war and above the routine, peaceful competition among states. It frequently involves protracted struggles of competing principles and ideologies. Low intensity conflict ranges from subversion to the use of armed force. It is waged by a combination of means employing political, economic, informational, and military instruments. Low-intensity conflicts are often localized, generally in the Third World, but contain regional and global security implications.

NOTE: Also known as small wars or "brushfire wars," low-intensity conflicts are rebellions, insurrections, and the like characterized by adversaries dedicated to a cause but lacking command and control structure and modern equipment.

Marine expeditionary force (MEF). The largest of the Marine air-ground task force (MAGTF) and the Marine Corps principal warfighting organization, particularly for larger crisis or contingencies. It is task-organized around a permanent command element and normally contains one or more Marine divisions, Marine aircraft wings, and Marine force service support groups. The Marine expeditionary force is capable of missions across the range of military operations, including amphibious assault and sustained operations ashore in any environment. It can operate from a sea base, a land base, or both. Also called MEF. See also aviation combat element; combat service support element; command element; Marine air-ground task force; Marine expeditionary force (forward); Marine expeditionary unit; special purpose Marine air-ground task force; task force. (JP 1-02)

maritime interception operation (MIO). Means by which an exclusion zone at sea is enforced. Involves the visit of merchant ships, by armed boarding parties if necessary, bound to, through, or out of a defined area, to examine each ship's papers and cargo and to search for evidence of contraband. The provisions must be established by a sanctioning body and must be applied to ships of all nationalities.

military operations other than war (MOOTW). Operations that encompass the use of military capabilities across the range of military operations short of war. These military actions can be applied to complement any combination of the other instruments of national power and occur before, during, and after war. (JP 1-02)

military occupational specialty (MOS). Primary job of military officer or enlisted personnel.

military operations in urban terrain (MOUT). All military actions planned and conducted on a topographical complex and its adjacent natural terrain where man-made construction is the dominant feature. It includes combat-in-cities, which is that portion of MOUT involving house-to-house and street-by-street fighting in towns and cities.

military police (MP). Marine Corps and Army police forces designated by specific MOS.

maritime pre-positioning force operation. A rapid deployment and assembly of a Marine expeditionary force in a secure area using a combination of strategic airlift and forward-deployed maritime pre-positioning ships. See also Marine expeditionary force; maritime pre-positioning ships (JP 1-02 and JP 4-01.6)

nonlethal weapons (NLW). Weapons that are explicitly designed and primarily employed so as to incapacitate personnel or materiel, while minimizing fatalities, permanent injury to personnel, and undesired damage to property and the environment.

a. Unlike conventional weapons that destroy their targets through blast, penetration, and fragmentation, nonlethal weapons employ means other than gross physical destruction to prevent the target from functioning.

b. Nonlethal weapons are intended to have one, or both, of the following characteristics:

(1) They have relatively reversible effects on personnel or materiel.

(2) They affect objects differently within their area of influence. (JP 1-02)

nonlethal weapon capability set. A set of systems containing weapon systems, munitions, and protective equipment. The components are explicitly designed and primarily employed to provide additional protection to friendly personnel and to incapacitate hostile personnel or material, while minimizing fatalities or permanent injury to intended targets and minimal damage to property and the environment. They are not required to have zero probability of producing fatalities or permanent injuries but are designed and employed in a manner that significantly reduces those probabilities over the traditional employment of currently fielded military weapon systems, munitions, and equipment.

nonlethal technology (N-LT). Synonymous with NLW, but also encompassing more advanced technology that is not yet available for use by military forces.

Oleoresin Capsicum (OC). Also known as "pepper spray", OC is made completely from organic materials and is FDA-approved for sale over-the-counter in the United States. Causes severe and immediate burning sensation to mucous membranes when sprayed into face, nose, and eyes.

psychological operations (PSYOP). Planned operations to convey selected information and indicators to foreign audiences to influence their emotions, motives, objective reasoning, and ultimately the behavior of foreign governments, organizations, groups, and individuals. The purpose of psychological operations is to induce or reinforce foreign attitudes and behavior favorable to the originator's objectives. (JP 1-02)

rules of engagement (ROE). Directives issued by competent military authority that delineate the circumstances and limitations under which U.S. forces will initiate and/or continue combat engagement with other forces encountered. See also law of war. (JP 1-02)

table of organization (T/O). Total manpower authorized for specific unit, organized by billet and rank. TO& E is a U.S. Army term which encompasses T/O and T/E.

training aids and devices. Any item developed or procured with the primary intent to assist in training and learning.

weapons of mass destruction (WMD). Weapons capable of a high order of destruction and/or of being used in such a manner as to destroy large numbers of people. Weapons of mass destruction can be high explosives or nuclear, biological, chemical, and radiological weapons, but exclude the means of transporting or propelling the weapon where such means is a separable and deliverable part of the weapon. (JP 1-02)

Index

Made in the USA
Monee, IL
05 November 2026

67808328R00213